Effective Customer Service

Ten Steps for Technical Professions

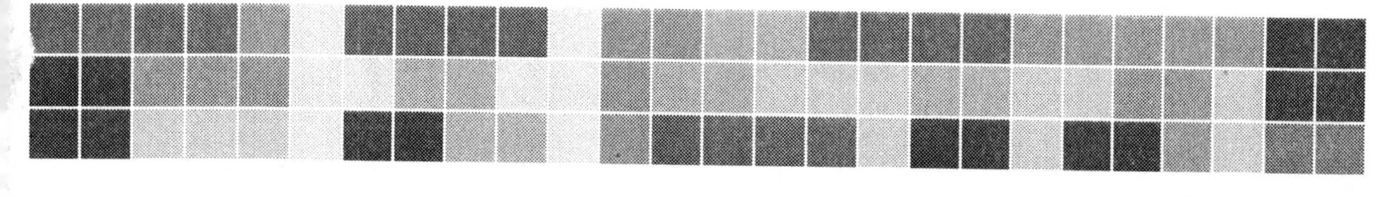

DAVID L. GOETSCH
CEO, Institute for Continual Improvement

STANLEY B. DAVIS
CEO, Stan Davis Consulting

NETEFFECT SERIES

Upper Saddle River, New Jersey
Columbus, Ohio

Editor in Chief: Stephen Helba
Executive Editor: Debbie Yarnell
Production Editor: Louise N. Sette
Production Supervision: Gay Pauley, Holcomb Hathaway
Design Coordinator: Diane Ernsberger
Cover Designer: Ali Mohrman
Production Manager: Brian Fox
Marketing Manager: Jimmy Stephens

Pearson Education Ltd.
Pearson Education Australia Pty. Limited
Pearson Education Singapore Pte. Ltd.
Pearson Education North Asia Ltd.
Pearson Education Canada, Ltd.
Pearson Educación de Mexico, S. A. de C.V.
Pearson Education—Japan
Pearson Education Malaysia Pte. Ltd.

ISBN 0-13-048529-2

Contents

About the Authors

David L. Goetsch is Provost of the Joint Campus of the University of West Florida and Okaloosa-Walton Community College and Professor of Quality, Management, and Safety. Dr. Goetsch is also President and CEO of the *Institute for Continual Improvement,* a private consulting firm dedicated to the continual improvement of employees, organizations, and communities. Dr. Goetsch welcomes feedback from his readers and may be contacted at the following email address: ddsg2001@Cox.net. **Stanley Davis** is Professor of Quality Management at Okaloosa-Walton Community College and CEO of *Stan Davis Consulting.*

Introduction

Ten-Step Model for Effective Customer Service

The concept of *globalization* brought many changes to the world of business. One of these was the realization that customers define quality. As this realization began to sink in on a broad scale, customer service became more important than it ever had been. It continues to be a critical issue for organizations attempting to survive and thrive in the competitive arena of global business. Few topics receive more attention in business literature than customer service; most of what is written approaches the subject from the perspective of the retail and hospitality industries, however. This is a serious shortcoming because effective customer service is just as important to other types of businesses as it is to restaurants, hotels, and retail establishments. For this reason, this book treats the subject of customer service from the perspective of engineering, manufacturing, construction, and other technology companies.

THE CUSTOMER IS KING

To prosper in today's competitive business environment, companies must understand that the customer holds the key to success. The customer must be at the very heart of the company's decision-making and thought processes. Said another way, in today's global marketplace, the customer is king. Consequently, the underlying philosophy of this book is that companies must become customer driven. The *customer-is-king* philosophy encompasses the following principles:

- *The customer must always be the company's highest priority.* The company's success, prosperity, and even survival depend on the customer.
- *Customers define value.* A company's products, services, and people are only as good as customers perceive them to be.
- *Loyal, lifelong customers are essential for long-term success.* Companies that consistently exceed customer expectations create loyal, lifelong customers.
- *Customer satisfaction is never permanent.* It is a fragile, fluid state that must be renewed continually.

HOW CUSTOMERS DEFINE VALUE—THE "VALUE EQUATION"

Companies that are determined to exceed customer expectations consistently over time must understand how customers define value. The value of a product or service is a combination of the customer's perceptions of the following factors. Each factor is one element in what the authors call "the value equation."

1. *Product quality.* The product must have the attributes customers want, and these attributes must meet or exceed customer expectations. Customer expectations do not relieve companies of their obligations to comply with applicable standards and regulations. In fact, compliance should be viewed as a given. A quality product complies with applicable standards and regulations, *and* it meets customer expectations.

2. *Service quality.* Providing a quality product is important, but that is just one piece of a larger customer-satisfaction puzzle. Companies must also ensure that customers are served promptly, efficiently, and courteously—whether their interaction is in person, by email, or on the telephone.

3. *People quality.* Customers build (or fail to build) relationships with employees in the company, particularly those they interact with frequently. An important ingredient in creating loyal customers is building and nurturing positive relationships between employees and customers. For this reason, it is important that customers be treated well by employees whose appearance, attitudes, and knowledge promote positive customer relationships. No matter how satisfied customers are with a company's products or services, they still might defect to the competition if they don't like the company's people.

4. *Image quality.* Image is not everything, but it is important. Image is important to companies because it is important to customers. Customers do

not want to be associated with companies that are bad corporate neighbors in their communities, have questionable reputations, or project a poor public image for any reason. Because of this, companies should concern themselves with both substance and image.

5. *Selling-price quality.* Selling price is always a critical factor in achieving customer satisfaction. Companies should assume that customers are sophisticated enough to know the difference between cheap and inexpensive. A quality selling price does not necessarily have to be the lowest price, but it must be competitive enough to ensure that potential customers consider the other elements of the value equation.

6. *Overall-cost quality.* Sophisticated customers consider not just the selling price but also the overall cost of a company's products or services. Companies that achieve the lowest selling price by hiding part of it in the overall cost soon alienate their customers—something companies cannot afford to do. Achieving quality in the area of overall cost requires that maintenance, upgrading, replacement parts, warranty issues, and after-purchase service are all factored in.

BECOMING A CUSTOMER-DRIVEN COMPANY

If the customer is king in today's marketplace, companies that hope to prosper must become customer driven. Being a customer-driven company means much more than adopting quaint slogans and putting up banners. Slogans and banners represent what the authors call the "cheerleading approach." All too often, companies that take the cheerleading approach display banners that say things like The Customer Is Always First, but they fail to do anything to actually put the customer first. Such an approach is a one-way street to bankruptcy. Customers are perceptive. If they are not treated well, if their business is not handled efficiently, and if they cannot count on the company to follow through on its promises, all the slogans and banners in the world contribute nothing to customer satisfaction. In fact, slogans and banners with no substance behind them are worse than no banners and slogans at all.

Becoming a customer-driven company requires commitment, time, patience, and effort. Customer-driven companies display the following characteristics:

- *Vision, commitment, and climate.* A company with these characteristics is totally committed to satisfying customer needs. Management demonstrates by deeds and words that the customer is important, that the organization is committed to customer satisfaction, and that customer needs take precedence over internal needs. One way organizations create a climate in which customer satisfaction is a priority is to make customer focus a major factor in promotions and pay increases.

▪ *Alignment with customers.* Customer-driven companies align themselves with their customers. Customers are included when anyone in the organization says "we." Alignment with customers manifests itself in several ways, including the following: customers play a consultative role in selling, customers are never promised more than can be delivered, employees understand what product attributes customers value most, and customer feedback and input are incorporated into the product development process.

▪ *Willingness to find and eliminate customers' problems.* Customer-driven companies work continually to identify and eliminate problems for customers. This willingness manifests itself in the following ways: customer complaints are monitored and analyzed; customer feedback is sought; and internal processes, procedures, and systems that do not create value for customers are identified and eliminated.

▪ *Use of customer information.* Customer-driven companies not only collect customer feedback, but they also ensure it is used by communicating it to those who can make improvements. The use of customer information manifests itself in the following ways: all employees know how the customer defines quality, employees at different levels are given opportunities to meet with customers, employees know who the "real" customer is, customers are given information that helps them develop realistic expectations, and employees and managers understand what customers want and expect.

▪ *Reaching out to customers.* Customer-driven companies reach out to their customers. In today's business environment, it is never enough to sit back and wait for customers to provide evaluative feedback. A competitive global marketplace demands a more assertive approach. Reaching out to customers means doing the following: making it easy for customers to do business, encouraging employees to go beyond the normal call of duty to please customers, attempting to resolve all customer complaints, and making it convenient and easy for customers to make their complaints known.

▪ *Competence, capability, and empowerment of people.* Employees are treated as competent, capable professionals and are empowered to use their judgment in satisfying customer needs. This means that all employees have a thorough understanding of the products and services they provide and of the customer's needs relating to those products. It also means that employees are given the resources and support needed to meet the customer's needs.

▪ *Continuous improvement of products and processes.* Customer-driven companies do what is necessary to continually improve their products and the processes that produce them. This approach to doing business manifests itself in the following ways: internal functional groups cooperate to reach shared goals, best practices in the business are studied (and implemented wherever they will result in improvements), research and development cycle time is continually reduced, problems are solved immediately, and investments are made in the development of innovative ideas.

TEN-STEP MODEL FOR EFFECTIVE CUSTOMER SERVICE

The previous section made the case for becoming a customer-driven company and explained several characteristics that set such companies apart. But how does a company develop these characteristics? How does a company become customer driven? This section sets forth a ten-step model for effective customer service. Each chapter in this book explains one of these steps in detail. The ten steps are as follows:

Step 1: Understand Effective Customer Service and Its Importance

Step 2: Set the Tone and Companywide Expectations

Step 3: Identify What Your Customers Want

Step 4: Benchmark the Company's Processes

Step 5: Compare Actual Performance Against Benchmarks, Identify Root Causes of Performance Problems, and Make Improvements

Step 6: Provide Training for Employees and Customers

Step 7: Turn Difficult and Dissatisfied Customers into Loyal, Repeat Customers

Step 8: Communicate Effectively and Often with Customers

Step 9: Establish Internal Customer Satisfaction

Step 10: Establish a Customer-Oriented Culture

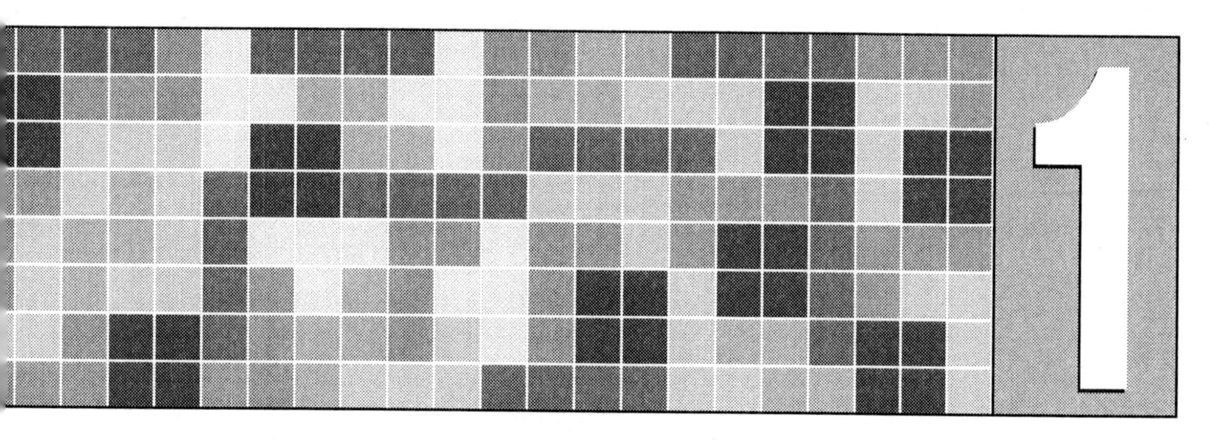

Understand Effective Customer Service and Its Importance

Too many companies operate on the leaky-bucket principle. They put all their effort into attracting more customers through marketing than they lose through neglect. In the long run, this approach just doesn't work.

GOALS

- Define effective customer service.
- Identify the goal of effective customer service.
- Name the characteristics of companies that practice effective customer service.
- Explain the rationale for effective customer service.

The first step in making effective customer service (ECS) a cultural reality is ensuring that everyone in an organization understands what ECS is and why it's important. In engineering, manufacturing, and construction firms, this means making sure that every employee—from the CEO to the lowest-paid worker—understands ECS and why it is important to the company. Consequently, this first step involves learning the definition, goal, characteristics, and rationale of ECS.

WHAT IS EFFECTIVE CUSTOMER SERVICE (ECS)?

Every person who has ever used the product or service of any organization knows the difference between poor customer service and ECS. Few people have actually tried to articulate what they intrinsically understand, however. We call it poor customer service when the people we interact with in a business are rude, have bad attitudes, are unwilling to help, seem uninterested, fail to acknowledge us, make us wait too long, or are condescending. We also call it poor customer service when they provide faulty, unreliable products that fail to meet our needs.

On the other hand, we tend to identify good customer service with companies whose employees are prompt, caring, knowledgeable, timely, courteous, flexible, and who stand behind products that are well-designed, well-made, reliable, and competitively priced. We call this type of customer service ECS, "effective" being defined as follows:

> *Services and products that exceed customer expectations consistently over the long run.*

This is a simple definition, but it is packed with substance. To fully appreciate the depth and significance of this definition, one must understand the following concepts, which are embedded in it: (1) to exceed customer expectations and (2) to maintain consistency over the long run.

Exceed Customer Expectations

It is never enough to simply meet customer expectations. The competition can usually meet the expectations of customers too. Consequently, ECS requires that companies go beyond meeting customer expectations to exceeding them, by providing more value than expected from the contract, transaction, interaction, and so on. Companies that thrive in today's competitive marketplace provide value to their customers. Value is a combination of customer perceptions of the following factors: product, service, people, image, selling price, and overall cost. None of these factors can be

Customer Value Equation

Value is determined by the customer's perception of the following factors:

	Product quality
+	Service quality
+	People quality
+	Image quality
+	Selling price
+	Overall cost
=	Value to the customers

FIGURE 1.1 Customers' perceptions of value in all of these areas are critical.

neglected. They are all important individually and, even more important, collectively (see Figure 1.1).

Maintain Consistency Over the Long Run

The goal of ECS is to create loyal customers who keep coming back and who make positive referrals to others. To achieve this goal, companies must exceed expectations consistently over time. Customer satisfaction is one of the most fleeting and fragile achievements in the world of business. Your customers are recruited unrelentingly by the competition. Every individual breakdown in customer service is an opportunity for the competition to get your customers' attention, and possibly their business.

THE GOAL OF ECS

The goal of ECS is this: *loyal customers for life.* By practicing ECS, your business attempts to create loyal customers who stay with you for life and who practice word-of-mouth advertising for your company. The latter element is important because word-of-mouth advertising by satisfied customers is the most effective advertising for any business. To create loyal, lifelong customers, it is necessary to understand what customer loyalty is—and also what it isn't. Such an understanding is important because certain customer behaviors can create the appearance of loyalty where it does not, in fact, exist; therefore, we discuss this first.

What Customer Loyalty Is Not

Companies should never mistake customer satisfaction at the moment for customer loyalty over time. Writing in the *Harvard Business Review*, Frederick F. Reichheld claims that between 65 and 85 percent of customers who defect to the competition say they were "satisfied" or "very satisfied."[1] Achieving customer satisfaction is a critical step toward achieving customer loyalty, but it is just one step. Even achieving a substantial share of the market in your industry is not a reliable indicator of customer loyalty. There can be many reasons that a company gains market dominance in its field, but unless the main reason is customer loyalty, the company's hold on its market is likely to be temporary.

Even repeat customers are not necessarily loyal customers. Some customers give your company repeat business for a period of time out of habit or familiarity rather than out of loyalty. There are two problems with this type of customer. First, they are prime targets for the competition. Habitual behavior and familiarity can both be overcome by competitors who are willing to work at it. Second, they are not likely to be good word-of-mouth advertisers for your business.

What Customer Loyalty Is

Customer loyalty is won and maintained by getting out in front of customers and seeking their input, rather than just their feedback, on all aspects of the value equation (product quality, service quality, people, image, selling price, and overall cost). It is important to understand the difference between "input" and "feedback." Input is given before the fact. In soliciting input, companies make customers into partners and learn about their preferences before providing the product or the service. Collecting customer input is about saying, "What do you need and how would you like it?" Customer feedback is collected after the fact and is about saying, "How did we do?" Companies should collect customer feedback, but should never rely on it alone. Instead they should collect both input and feedback.

If satisfaction at the moment, repeat business, and market share are not reliable indicators of customer loyalty, how can a company know who its loyal customers are? Loyal customers are recognizable by several behaviors and characteristics, including those discussed below. Figure 1.2 summarizes these points.

1. *Willingness to participate in activities designed to collect customer input and feedback (e.g., focus groups, surveys, critique sessions).* Providing input and

Characteristics of Loyal Customers

- Willing to provide frank, honest input about all aspects of the value equation.
  - Participate in focus groups
  - Complete surveys
  - Participate in critique sessions
- Willing to work with representatives of the company to establish and continually improve relationships.
- Willing to recommend the company to potential customers.
  - Do word-of-mouth advertising
  - Introduce company representatives to potential customers
- Willing to resist the recruitment efforts of the competition.
  - Inform the company of recruitment efforts
  - Give the company an opportunity to match a better offer

FIGURE 1.2 How to recognize loyal customers.

feedback takes time, thought, and effort. Customers who are willing to participate show they are willing to invest in a long-term relationship with your company. An investment is a commitment, and a commitment is the foundation of a loyal, lifelong customer relationship.

2. *Willingness to work with the company and its representatives to nurture and continually improve the relationship.* People are people no matter what kind of business is being conducted. A predictable characteristic of people is that they like to do business with others they can relate to, respect, and—ideally—like. Consequently, a customer's willingness to cooperate in efforts to build relationships with representatives of the company is a strong indicator of customer loyalty.

3. *Willingness to recommend the company to potential customers.* No other kind of advertising is as effective as word-of-mouth advertising by a satisfied customer. When a satisfied customer recommends your company to a potential customer (prospect), the recommendation has more credibility than the most cleverly designed advertising campaign or the most talented sales representative. Customers who recommend your company to prospects are expressing what is perhaps the most sincere and valuable form of loyalty.

4. *A demonstrated resistance to the recruitment efforts of the competition.* Every one of your customers is the target of aggressive competitors every day. It is not uncommon for a competitor to offer a better selling price in an attempt to lure your customers away. When this happens, loyal customers respond by giving your company the benefit of the doubt and looking at all elements of the value equation (see Figure 1.1). If, by chance, the value equation favors the competitor, a loyal customer will come to you and lay out the facts, giving your company an opportunity to match the deal. In addition, a loyal customer factors in the "longevity element" after considering all aspects of the value equation. This means that even if a competitor gains a slight advantage in the value equation, your company's proven record over time will give you the edge.

CHARACTERISTICS OF COMPANIES THAT PRACTICE ECS

Companies that practice ECS share a number of common characteristics. Those described in this section can be used by any company as a self-test and as a framework for collecting customer feedback. Customer feedback about your company's performance with regard to these characteristics should be an important element in ongoing self-assessments.

Eager to Solve Customer Problems

When customers have problems, companies that practice ECS go out of their way to help solve them. Problems are not ignored or handed off to someone else. Rather, they are "adopted" by the first company representative who becomes aware of the problem, regardless of this person's position in the company. Even if an employee does nothing more than connect the customer with the right people, she has performed a valuable service to the customer *and* to the company. Consider the following examples, one bad and one good. First, the bad example. A customer calls your company's engineering department to say he cannot download certain files an engineer sent him electronically, and a deadline is approaching fast. The telephone is answered by another engineer who says, "He's in a meeting right now. Call him back in an hour or so." Now consider the good example. The scenario is the same as before, but this time the person who answers the telephone says, "That engineer is in a meeting that could last all morning, but I'll go online with you right now and see if I can talk you through the problem. If we can't figure out how to download the files, I'll pull your contact out of his meeting. Don't worry, we'll get this problem solved in time to beat your deadline."

Take a Long-Term View of Customer Relationships

Companies that practice ECS approach customer relationships from a long-term perspective. Their personnel are trained to consider what is best for the long-term good of the relationship, and are expected to apply this training every time they interact with a customer. This means that "quick fixes" to problems, which will just cause even worse difficulties later, are not an option. When a quick fix is necessary as a temporary expedient, you and the customer should agree on this approach and establish a mutually satisfactory timetable for implementing a permanent solution.

Accept Responsibility for Correcting Problems

Companies that practice ECS take responsibility for correcting problems *even when customer actions contribute to them*. There was a time when the most popular customer-service slogan was, "The customer is always right." One of the reasons this slogan fell out of favor is that it is so obviously false. Quite often when problems arise with a company's product, not only is the customer wrong, but her actions actually contributed to the problem. Improper use of a product by a customer is common. Even when this is the case, however, ECS companies take responsibility for solving the customer's problem. Notice that responsibility is taken for solving the problem, not for causing it. Blaming the customer for failing to read the instructions or for other contributory behavior might make you feel better, but it will do nothing to promote loyalty in the customer. If a customer misuses a product and claims to have read the instructions, it is smart to assess the adequacy of the instructions provided with the product. When doing so, get some feedback from a third party. Maybe your instructions are poorly written or overly complicated.

Provide an Unexpected Benefit to the Customer

In every transaction in which a product or service is provided to a customer, ECS companies provide some added benefit the customer will appreciate, but does not expect. A construction company that specializes in large office buildings may give the owner a scale model of the building, completely landscaped, in a special case for display in the lobby of the building. A manufacturer may give a customer an extra six months on its product warranty. An engineering firm may install some newly purchased equipment at no charge to the customer. These extras, although small when compared with the monetary value of the transaction in question, can be important in winning the customer's loyalty. The value of these little extras can be enhanced significantly by conducting some quiet research into the types of things a given customer would really like.

Value the Time of Customers

Companies that practice ECS know that a customer's time is valuable, and they conduct their business accordingly. This means training all personnel to return telephone calls and to respond to email messages promptly. It means being not just on time, but a little early for all appointments with customers. It means being well prepared for every customer interaction so that time is not wasted looking for information or material needed to transact the customer's business or solve the customer's problem. It also means being cognizant of the time of day at which customer interactions are initiated. Calling a customer five minutes before the end of the business day is both inconsiderate and ill advised. Management personnel set a positive example by practicing these behaviors when interacting with employees as well as with customers. In fact, a good rule of thumb for management personnel is: Treat your employees the way you want them to treat your customers.

Participate in Ceremonial Activities Organized by Customers

Certain milestone business transactions are recognized with an appropriate ceremony. The completion of a new building, the one-millionth barrel of a given substance purchased, the signing of a large contract, the unveiling of the prototype of a new product—these are all examples of milestones that might be recognized with a ceremony. When the customer is the party organizing the ceremony, ECS companies not only participate; they also help out where possible. Customers should feel that their ceremony is as important to your company as it is to theirs.

Make Customers Feel Welcome

Making customers feel welcome is just as important for engineering, manufacturing, and construction companies as it is for restaurants and retail establishments. ECS companies are cognizant of how customers are treated when they visit the company in person, online, or by telephone. Does your company have a comfortable place for customers to wait when they arrive early for a meeting? Are the customers' needs and preferences considered when setting up meetings (i.e., refreshments, meals, and so on)? Are customers greeted enthusiastically by all employees and not just the receptionist who has been trained in this area? What type of greeting do customers get when they call your company? Is your company's webpage customer friendly, and does it have a welcome feel to it? ECS companies can give affirmative answers to questions such as these.

Stand Behind Their Products and Services

ECS companies provide satisfaction guarantees that show they stand behind their products and services. These guarantees can be stated explicitly in writing or implicitly through the company's actions. In the long run, the latter approach is the more effective. Customers know by your company's actions whether or not you stand behind your products and services. If you don't, all the written guarantees in the world do not matter. Also, written guarantees cleverly crafted with loopholes that allow your company to wriggle out of its responsibilities to the customer will do more harm than providing no guarantees at all.

Expect, Model, Monitor, Evaluate, and Reward ECS

ECS cannot be just a program, a campaign, or an initiative. It must become a way of life. Companies that practice ECS embed the concept in their performance infrastructure by expecting, monitoring, modeling, and rewarding ECS behaviors. Expectations regarding interactions with customers are laid out in employee job descriptions. In addition, management and supervisory personnel communicate constantly about the expected behaviors. What is expected of employees is modeled by management personnel. This is important.

ECS companies know that employees tend to treat customers the way they are treated by their managers. When management personnel model the behaviors they expect in their employees, peer pressure among employees tends to support the desired behavior. The obverse is also true: If employees observe managers saying one thing but doing another, peer pressure among employees works against the desired behavior. Management personnel monitor employees constantly to ensure that expectations are being met, and they encourage employees to monitor each other. Employees at all levels are encouraged to speak up, in private, when they see behavior that falls short of the company's expectations. This establishes positive peer pressure that supports the desired behaviors. Not only are customer-service behaviors monitored, they are evaluated. Performance appraisal instruments include criteria related to the customer-service behaviors contained in job descriptions. In this way, customer-service performance becomes one of the company's criteria for promotions and rewards. Figure 1.3 is a self-assessment instrument companies can use to determine if they have the characteristics of businesses that practice ECS.

THE RATIONALE FOR ECS

The rationale for ECS is best explained using a simple logic chain, as follows: (1) companies need loyal customers to survive and succeed; (2) customers are loyal to companies that best meet their needs in terms of the

ECS Characteristics Self-Assessment Instrument

In completing this self-assessment instrument, you are encouraged to seek customer feedback when applicable.

Yes	No	Criteria
		Are all employees eager to help customers solve problems?
		Do employees take responsibility for helping customers even when the problem is outside of their area?
		Do employees work with customers to find permanent solutions to problems rather than settling for temporary "quick fixes"?
		Do employees accept responsibility for helping customers solve problems even when the customer's actions contributed to the problem?
		Does your company provide an unexpected but appreciated benefit to customers at the conclusion of a contract, delivery, or transaction?
		Do employees value customers' time?
		Does your company participate in ceremonial activities organized by customers?
		Does your company have a "welcome feel" to customers?
		Does your company facilitate access when customers must visit restricted access areas?
		Does your company have a comfortable place for customers to wait when they show up early for meetings?
		Are the customer's needs and preferences for refreshments and meals considered when setting up meetings?
		Are customers greeted enthusiastically by all employees?
		Do customers receive positive, friendly treatment when they call on the telephone?
		Does your company's webpage have a welcome feel to it?
		Does your company stand behind its products and services?
		Are customer service expectations explained in employee job descriptions?
		Has your company adopted and communicated a companywide, comprehensive statement of its customer-service philosophy?
		Do management and supervisory personnel monitor the customer service behavior of their employees and each other?
		Do employees monitor each other in terms of customer-service behavior?
		Do your company's performance appraisal instruments contain customer-service criteria?
		Do management and supervisory personnel consider customer-service behaviors when making decisions about employee raises, promotions, and other rewards?

FIGURE 1.3 Companies should conduct periodic self-assessments of their customer-service characteristics.

value equation; and (3) companies that practice ECS effectively perform better in terms of the value equation than do other companies. Therefore, companies should practice ECS. In addition to the logic just presented, there are other good reasons for practicing ECS, including preventing the predictable behavior of dissatisfied customers and avoiding the cost of lost customers.

Predictable Behavior of Dissatisfied Customers

Dissatisfied customers behave in predictable ways that are bad for companies that must compete to survive. This section contains information that is widely known in customer-service circles about the behavior of dissatisfied customers.[2]

1. Dissatisfied customers often respond by talking. They tell between 8 and 10 people of their negative experience with your company.
2. Only about 30 percent of dissatisfied customers extend the courtesy of telling you about their problem—but they will tell others.
3. Almost 5 percent of the people who want to complain go directly to the headquarters level.
4. When dissatisfied customers are unhappy about how their complaint was handled, almost 70 percent will refuse to do business with your company again.
5. It typically costs a company 10 times more to attract new customers than to retain current customers.

These facts reveal the tendency of dissatisfied customers to become former customers. In addition and maybe worse, they spread their dissatisfaction among prospective customers. Negative word-of-mouth advertising tends to spread even faster than the positive variety. This can do irreparable damage to a company's reputation in a very short period of time. Just these few facts show the critical nature of customer satisfaction.

Cost of Lost Customers

One lost customer is, in reality, more than just one. If losing a customer meant nothing more than losing just one contract, customer satisfaction would be less of an issue. Most companies can overcome the loss of one customer. But this is not how lost contracts work. They are never just isolated incidents. When computing the cost of lost customers, consider all of the following factors:

1. Cost of the one lost contract in question
2. Cost of future contracts with this same customer

3. Cost of the "ripple effect" as this dissatisfied customer tells 8 to 10 prospective customers
4. Cost of replacing the lost customer (typically at least 10 times more than retaining an existing customer)

Summary

1. Effective customer service, or ECS, is defined as services and products that exceed customer expectations consistently over the long run. ECS requires that companies go beyond meeting customer expectations to exceeding them by providing more value than they expect from the transaction, contract, or interaction. Value is a combination of customer perceptions of the following factors: product, service, people, image, selling price, and overall cost. Customer satisfaction is fragile and fleeting. Consequently, ECS is aimed at exceeding customer expectations not just once, but consistently over time.

2. The goal of effective customer service is creating loyal customers for life. Customer satisfaction is a critical step in the process of winning customer loyalty, but it is just a step. Customer satisfaction alone does not guarantee customer loyalty. Even marketshare is not a reliable indicator of customer loyalty. There can be many reasons that a company gains market dominance, but unless the main reason is customer loyalty, the company's hold on its market is likely to be temporary. Surprisingly, repeat customers may not be loyal customers. Repeat business can be attributed to habit or familiarity rather than loyalty. Customer loyalty is won and maintained by asking customers for their input and feedback on all elements of the customer value equation.

3. Loyal customers can be recognized by several characteristics, including the following: (a) willingness to participate in activities designed to collect customer input and feedback; (b) willingness to work with the company and its representatives to nurture and continually improve the relationship; (c) willingness to recommend the company to potential customers; and (d) a demonstrated resistance to the recruitment efforts of the competition.

4. Characteristics of companies that practice ECS include the following: (a) eager to solve customer problems; (b) take a long-term view of customer relationships; (c) accept responsibility for correcting problems, even when the customer's actions contribute to them; (d) provide an unexpected benefit to the customer; (e) value the time of customers; (f) participate in ceremonial activities organized by customers; (g) make customers feel welcome; (h) stand behind their products and services; and (i) expect, model, monitor, and reward ECS.

5. The rationale for ECS is that it represents the best way to win and maintain loyal customers in a competitive environment. Dissatisfied cus-

tomers behave predictably—they migrate to the competition, and they spread the word about their dissatisfaction. A lost customer usually represents more than just one loss. Companies must also factor in the cost of future business the lost customer might have brought in, the ripple effect when the lost customer talks with 8 to 10 potential customers, and the cost of replacing the lost customer (which can be 10 or more times higher than the cost of retaining a customer).

Key Phrases and Concepts

Caring, knowledgeable people
Competitive overall price
Competitive selling price
Consistency over the long run
Customer feedback
Customer input
Customer value equation
Eager to solve customer relationships
Effective customer service (ECS)
Exceed customer expectations
Expect, model, monitor, evaluate, and reward
Loyal customers for life
Make customers feel welcome
Participate in ceremonial activities
Positive image
Resistance to recruitment efforts
Stand behind products and services
Unexpected benefit to the customer
Willingness to recommend

Review Questions

1. What is effective customer service (ECS)?
2. Explain the customer value equation.
3. Explain the concept of exceeding customer expectations and why this is important.
4. Explain the concept of exceeding customer expectations consistently over the long run.
5. What is the goal of ECS?
6. Explain why customer satisfaction is not enough to ensure customer loyalty.
7. Explain why marketshare is not a reliable indicator of customer loyalty.
8. Explain why repeat customers may not necessarily be loyal customers.
9. How can a company recognize loyal customers?
10. List and explain the characteristics of companies that practice ECS.

11. What is the rationale for ECS?
12. Describe the predictable behaviors of dissatisfied customers.
13. What factors should be considered when estimating the cost of a lost customer?

ECS APPLIED: DIVERSIFIED TECHNOLOGIES COMPANY BEGINS IMPLEMENTATION OF ECS

This is the first installment of a serialized case presented over the course of this book, one installment per chapter. The overall case shows how one technology company that has engineering, manufacturing, and construction divisions came to the realization that it needed to adopt ECS, and how it went about doing so. The challenges presented by this change, and the way the company met the challenges, are typical of the experience of most organizations that adopt the ECS approach to doing business. Each installment in this serialized case and the case taken as a whole should be instructive for any company.

Diversified Technologies Company (DTC) is a medium-sized firm (1,200 employees) with engineering, manufacturing, and constructions divisions. Its services are comprehensive and include mechanical, electrical (analog), electronic (digital), civil, structural, and computer engineering, as well as manufacturing and construction services. Principal officers of the company are CEO David Stanley, Vice President for Engineering Meg Stanfield, Vice President for Manufacturing Tim Wang, and Vice President for Construction Conley Parrish. DTC has been in business for more than 30 years and has both domestic and international customers. At its peak, DTC did $120 million in business. Over the past three years, however, the company has experienced a slow but steady decline in business of about 4 percent. To stop the decline, turn the business around, and get back into growth mode, David Stanley is considering adoption of the ECS approach to doing business. Stanley called a meeting of his executive management team (himself and the three vice presidents) to discuss ECS.

"We are losing marketshare, and the only thing I've been able to tie the problem to is competition," said the CEO. "It's getting tougher and tougher to hold on to our customers. Last week we lost DaraTech to a competitor in Germany. I can remember recruiting DaraTech myself back when I was running the Engineering Division. Don James was my contact. He's their CEO now. I never thought I'd see the day we would lose DaraTech. Until last week they'd been with us for 15 years." Stanley then told the vice presidents about an article he had read on a concept called effective customer service, or ECS. "I think this might be the answer we've been looking for."

"Sounds like just another three-letter acronym to me," said Tim Wang. "Those things are a dime a dozen and have the shelf life of cottage cheese." "I agree," said Conley Parrish. "I'll bet if I tried I could write down at least 20 acronyms that some consultant dreamed up to sell books and videotapes over the past 10 years, all of them claiming to be miracle cures for what ails business." "I don't know," said Meg Stanfield. "I don't like the acronyms either, but some of the ideas that come out of these concepts are worthwhile. You didn't like TQM either Tim, but now you're the world's biggest advocate of continual improvement." "I hate to admit it," said Tim Wang, "but Meg makes a valid point. Tell us about this latest acronym, ERS or XYZ or whatever it is." "ECS," said Stanley. "It stands for effective customer service. But forget the acronym for the moment and let's look at the concept. I think it's worth considering. It's not a miracle cure or anything like that and doesn't claim to be, but it is a different approach to doing business. And we need a different approach. For whatever reason, what we're doing now isn't working."

"I think the best way to explain ECS is to ask a series of questions," said the CEO. He then proceeded to ask his vice presidents a number of questions, adding comments to facilitate discussion. "Are our customers willing to help us out when we want customer input or feedback?" Meg Stanfield commented that her division had never asked customers for input, but she certainly got plenty of feedback when they weren't happy. Conley Parrish's division had never asked for input either. Tim Wang had attempted to organize formal input sessions as a part of his continual improvement efforts, but without success. "Let me ask all of you another question," said Stanley. "Do you know of any new customers that have come to us based on the recommendation of an existing customer?" The vice presidents couldn't name one.

"Are our customers willing to work with us to improve relationships between our people and theirs?" asked the CEO. Tim Wang spoke for the group when he said, "I don't know how to answer that. We've never asked them to do anything like that." Stanley's last question was, "Have our customers ever demonstrated resistance to the recruitment efforts of our competition?" Then, before the vice presidents could answer he said, "Never mind. The answer to that question is self-evident."

"Here is what I learned from the article I read about ECS." Stanley told the vice presidents that companies couldn't survive in a competitive environment without loyal customers. Then he read them the characteristics of loyal customers. "Obviously, we don't have loyal customers, and to be frank, it's our fault." Stanley then gave each of the vice presidents a copy of the article he had read on ECS. He had highlighted the section of the article that presented a ten-step process for implementing the concept. After giving the vice presidents time to look over that portion of the article, Stanley asked, "Does anyone see any reason why we shouldn't give ECS a try?" When he got no responses, Stanley asked the vice presidents to get together again in a week to develop a plan for communicating all the information in the ECS article to all employees in the company.

Within a month, every employee at DTS knew about ECS and why it is important. The company had taken the first step in the implementation process.

DISCUSSION CASES

The following cases provide examples of how the various concepts presented in this chapter might play out in actual companies. The cases are provided to prompt discussion, give the reader a feel for the types of problems confronted in the workplace, and reinforce the ECS concept in question.

CASE 1.1 Failure to Understand the Importance of Customer Service

At one time, Tycon, Inc. was a major success story in the field of electronics manufacturing. Its founder and CEO, Jack Adams, had worked in the electronics manufacturing business for 20 years before deciding to go it alone. He started the company in his garage and was its only employee during Tycon's first year of business. But the electronics industry was booming at the time, and it wasn't long before Adams had Tycon in a rapid-growth mode. Within just six years, Tycon grew from a one-man show into a thriving manufacturing company employing more than 500 people.

Adams made Tycon competitive by locating it in a military community. This served several important purposes. First, this put him in close proximity to the local military base, a major customer for the larger defense contractors that represented Tycon's principal customer base. Second, and probably more important, it allowed Tycon to hire retired military personnel, who brought applicable experience, excellent skills, and a positive work ethic to the job. Third, the base ensured Tycon a steady supply of outstanding employees who were willing, because of their military retirement pay, to work for wages below the industry average. This formula kept Adams and Tycon on top in the electronics manufacturing business for many years. However, over time Tycon's business began to taper off. Then, almost as if it had happened overnight, the company's business began to plummet alarmingly.

The larger defense contractors Tycon depended on for work originally were attracted to the company because its bid packages were well prepared, thorough, and clearly at the low end of the bidding spectrum. When a low bid was the issue, Tycon was usually the winner. However, when constant cost overruns on military contracts became an issue nationwide, defense contractors began to focus not just on the lowest bid (selling price), but also on the overall cost of products and services. When this new focus became the norm, Tycon's business dropped off almost as fast as it had grown in the early days.

In danger of losing his company, Adams brought in a team of consultants to review Tycon's business practices from top to bottom and make recommendations. Their report could have been summarized in just three words: "poor customer service." Adams and his key decision makers at Tycon had focused so intently on being the lowest bidder that they had paid too little attention to product quality, service, image, and overall cost. When the rules of the game in defense contracting changed, Tycon's business practices didn't. Eventually Adams was able to turn things around at Tycon by focusing his team on effective customer service, but not before seeing his company shrink from more than 500 employees to fewer than 100. Adams and Tycon are climbing back up the ladder, but it's a slow process, and the competition is intense. Adams found out the hard way that one of the most difficult things to change in business is a bad image.

Discussion Questions

1. Have you ever had a bad experience with a company that damaged the image you had held of that company? If so, explain your experience.
2. What would this company have to do to regain your confidence?

CASE 1.2 We Are the Market Leader—Why Worry?

Excel Manufacturing, Inc. (EMI) was founded by Don Anchort, CEO, and Mack Tidwell, vice president for engineering and manufacturing. EMI is now an effective practitioner of ECS, but this was not always the case. The company decided to adopt ECS during a meeting of its executive management team several years ago, but not before Anchort and Tidwell had come dangerously close to parting ways over the issue of customer service.

Anchort and Tidwell had started the company while still in college studying engineering. This was in the early days of the personal computer when random access memory was still talked about in terms of kilobytes. They were a perfect team from the outset. Anchort was a natural inventor and designer. Tidwell, on the other hand, could make anything Anchort designed. They both envisioned a need for what they still call "shells," the plastic and metal consoles that hold the printed circuit boards and other electronic mechanisms for personal computer towers, printers, and display terminals. As the computer business took off globally, so did their fledgling company. Within just five years of its founding, EMI was the largest private-sector employer in its community.

After just 10 years in operation, EMI was a market leader in its field. This pleased Anchort, but it scared Tidwell. So began a debate between the company's two founders that got heated before it got better. From Anchort's perspective, EMI should keep doing what it had always done and in the way it had always done it. His view was simple. "We are on top, so why do anything different? You don't fix what isn't broke." Tidwell disagreed. He too appreciated the ideas behind the company's rapid growth, but he was convinced that standing on top of the mountain just made EMI an easier target for its competitors. His view was one of caution. "If we don't get focused on customers, some other company is going to sneak up behind us one day and kick EMI off this mountain."

The debate between the two partners went on for more than a year before it came to a head. After a heated argument that took place while their two families were sharing what should have been a happy Christmas dinner, Anchort and Tidwell decided they had to put the issue to rest once and for all. After the Christmas holiday, Anchort called a meeting of EMI's executive management team and set the argument before them without revealing which thoughts were his and which were Tidwell's. He simply laid out the two sides of the debate, and then asked the other members of the team for their thoughts. Although he did a good job of hiding it, Anchort was surprised to learn that the company's other executives supported Tidwell's side of the argument (without knowing it represented Tidwell's sentiments). The clincher in changing Anchort's mind was a statement made by the company's chief financial officer, who said, "Right now, somewhere in the world, there are two young engineering students in a garage plotting ways to do what we do and do it better. They are just as smart as we are, probably smarter. Plus, we have plowed the ground for them in terms of both design and manufacturing processes. If we want to stay on top, we have to keep thinking, keep innovating, and keep improving. But just as important, we have to focus on our customers. We need to make them partners in our creative processes, ask them how we are doing, learn what we need to do better, and make sure they know how much we appreciate their business. This is how we will stay on top."

Within a week of this meeting, Anchort and Tidwell had started EMI down a path toward full adoption of the ECS approach to doing business. Now, several years later, Anchort is a strong and articulate advocate of the concept. At last year's Anchort/Tidwell Christmas dinner, there was a special guest: the CEO of EMI's oldest and largest customer, a customer that had begun talking to competitors before EMI finally asked for its input—and acted on it.

Discussion Questions

1. Have you ever dealt with a company that appeared to take your business for granted? How did you feel about this company?
2. Would you stop doing business with a company that took your loyalty for granted?

CASE 1.3 Customer Perceptions Can Be Surprising

Angela Wong's engineering services company—Wong & Associates—was small, just 30 employees. But Wong always took great pride in saying, "We may not be the biggest, but we are the best." This was how she marketed her little company, and the results had been encouraging. In spite of her success, Wong was a realist. She knew that the better Wong & Associates served its customers, the more her company would grow. Wong welcomed the growth, but she did not want to lose what she called the "personal touch" with customers. This was why she instituted a customer feedback program run by a third-party facilitator to ensure openness and objectivity, a program that provided Wong with some interesting information about customer perceptions of her company. Wong learned from the results of the first customer focus group organized by her facilitator that the self-perceptions of Wong & Associates personnel and the perceptions of customers didn't match.

In reading the facilitator's report, Wong was surprised to learn that customers don't think the company was as responsive as it should be when problems arise. One long-term customer said, "In the early days, if I called, I talked directly to Angela and she solved my problem personally right then. Now I usually talk to one of her engineers. Most people I talk to at Wong & Associates are OK about helping me, but they are not as eager as Angela always was." This comment cut Wong deeply, because it showed that her worst fear, the fear of losing the personal touch, was actually coming to pass as the company grew. She also learned she had problems with some employees who did not value customers' time as they should. There were instances noted by customers of Wong & Associates personnel taking too long to return calls, occasionally showing up for meetings unprepared, and holding customers up by arriving late for meetings.

Within a week of reading the facilitator's report, Angela Wong called a companywide meeting and spoke with all her employees. During this meeting, she made sure every employee understood her customer-service expectations. She also informed employees that their job descriptions and performance appraisals would be updated to include these expectations. "Being a good engineer, CAD technician, or secretary is critical, but it is no longer enough," said Wong to her employees. "You also need to be good customer-service representatives for our company. From now on, your bonuses, raises, and promotions will depend on it. Starting next week, we are all going to attend a customer-service training program held here in our conference room every Monday afternoon for a month."

By the time the next semi-annual customer focus group had taken place, Wong & Associates had regained the personal touch Angela Wong had used to grow the company. Wong & Associates now had all of the characteristics of a ECS company, and its employees were committed to continually improving on each of the characteristics.

Discussion Questions

1. Have you ever dealt with a company that seemed uninterested in your problems? That was unresponsive? Explain.
2. What would such a company have to do to win your confidence?

CASE 1.4 The Company Should Have Solved the Problem and Kept the Customer

Maxwell Engineering Company (MEC) provides civil, structural, electrical, and mechanical engineering services to architects and construction companies. For months, MEC's President had worked hard to attract the architectural firm of Baker & Cobb as a customer. B&C is an established firm that specializes in urban redevelopment. With B&C as a satisfied customer, MEC could look forward to a substantial increase in its workload. This is why MEC's behavior toward B&C was so difficult to understand. Through a mix-up in B&C's architectural plans, MEC's corresponding electrical plans caused the electrical contractor to improperly wire the third and fourth floors of a high-rise office building. B&C's personnel admitted that their architectural drawings were partially to blame in the matter, but also felt that some common-sense communication from MEC's engineers could have prevented the mix-up. Consequently, B&C's chief architect asked MEC to split the cost of rewiring the floors in question. MEC's vice president for marketing recommended approval of B&C's plan to split the costs. However, MEC's president, known in the industry as a tough businessman, refused. He claimed that B&C was fully to blame and that, as a result, it should pay the entire cost of rewiring the two floors. Seeing no other alternative, B&C's personnel capitulated and paid. MEC saved some money in the short run, but at a severe cost in the long run. The architect's CEO, unhappy with MEC's stubborn stance, told his personnel to scratch MEC from its list of engineering partners. This was done, and MEC's president was informed, but the architects didn't stop there. In addition, B&C personnel must have talked to their counterparts in other architectural and construction firms, because over the next three years MEC saw its business decrease—during a period in which it should have grown. MEC's president learned the hard way that in a competitive marketplace you sometimes have to give in order to get.

Discussion Questions

1. Have you ever stopped doing business with a company because it failed to treat you properly? Explain.
2. What would such a company have to do to win you back?

Endnotes

[1]Fredrick F. Reichheld, "Loyalty-Based Management," *Harvard Business Review* (March–April 1993), p. 71.

[2]Richard C. Whitely, *The Customer Driven Company: Moving From Talk to Action* (New York: Addison-Wesley, 1991), pp. 221–225.

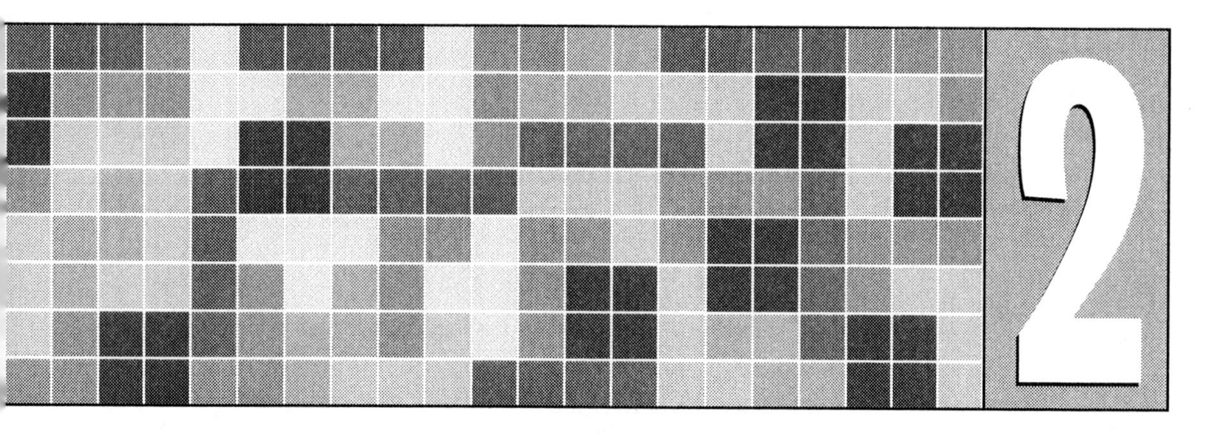

Set the Tone and Companywide Expectations

Employees learn what is important from the actions of management. If managers want employees to value customer service, they have to set the right tone and the right example.

GOALS

- Learn how to integrate customer service into the company's strategic plan.
- Learn how to integrate customer service into all job descriptions.
- Learn how to integrate customer service into the performance-appraisal process.
- Learn how to integrate customer service into the company's reward and recognition systems.
- Develop a customer-service philosophy statement.
- Set a positive example.

Components of a Company's Employee-Performance Infrastructure

- ✓ Strategic plan
- ✓ Job description
- ✓ Performance-appraisal process
- ✓ Reward and recognition system
- ✓ Customer-service philosophy statement
- ✓ Positive role models

FIGURE 2.1 The various elements of the infrastructure support the type of employee performance desired.

Employees follow the lead of management in deciding what is important. Consequently, it is critical that managers set the right tone, set a positive example, and establish high expectations for customer service. In attempting to set the tone, it is important that managers' actions match their words. The *actions* of managers are even more important than their words. Words can be used to augment actions, but they can never replace actions. For this reason, we recommend against high-profile customer service "campaigns" that consist of a lot of hoopla and cheerleading. Slogans and banners can be useful, but they should be employed only after the company institutes its employee-performance infrastructure to support customer service. The components of this infrastructure are summarized in Figure 2.1 and discussed below.

THE COMPANY'S STRATEGIC PLAN

Competitive companies adopt strategies to maximize their performance in the marketplace. These strategies are typically set forth in a comprehensive document called the "strategic plan." A comprehensive strategic plan has the following components: a vision statement, a mission statement, guiding principles, and broad strategic goals. Expectations regarding customers should be reflected in all of these components.

The Vision Statement

The company's vision statement should reflect the dream of what a company would like to be. It should be a beacon in the distance toward which

Characteristics of a Well-Written Vision Statement

- Customer-oriented
- Easily understood by all stakeholders
- Challenging, yet attainable
- Lofty, yet tangible
- Capable of stirring excitement for all stakeholders
- Capable of creating unity of purpose among all stakeholders
- Not concerned with numbers
- Sets the tone for employees

FIGURE 2.2 A company's vision statement should have these characteristics.

the company always strives. Vision statements should have the characteristics summarized in Figure 2.2.

"Customer oriented" is the critical characteristic of the vision statement. Each of the characteristics is important, but the customer-orientation characteristic is the first priority. Here are examples of vision statements:

- Morgan Engineering, Inc. will be recognized by its customers as the provider of choice for engineering services in the "tri-state" region.
- TecMan Corporation will be recognized by its customers as the leading manufacturer of color copiers in the United States.
- Daniels Construction Company will be recognized by its customers as the premiere builder of multistory parking garages in the southeastern United States.

These examples are all well-written vision statements that contain the recommended characteristics. Notice that in all three statements, the recognition sought is that of customers. In a competitive environment, it matters little if these companies recognize themselves as the best or even if third-party certification agencies recognize them as the best. What does matter is that *customers* recognize them as the best. Self-recognition is irrelevant because companies don't purchase their own products or services. Recognition by third-party organizations that award certifications (e.g., ISO 9000, ISO 14000, the Baldrige Award) are good. But it must be remembered that third-party organizations don't buy a company's products or services either. Recognition by entities other than customers has competitive value only to the extent that it helps a company gain the desired recognition of its customers.

The Mission Statement

The mission statement tells why a company exists. Whereas the vision statement is a qualitative expression of what the company would like to be, the mission statement is a practical expression of what the company is. Here are examples of mission statements:

- Morgan Engineering, Inc. provides high-quality engineering services to customers in the following disciplines: mechanical, electrical, industrial, and civil engineering.
- TecMan Corporation is an electro-mechanical manufacturing company that provides high-quality color printers for business, industry, and government customers.
- Daniels Construction Company specializes in the construction of multistory parking garages for customers in urban settings in the southeastern United States.

These statements explain why the companies exist, what kinds of products or services they provide, and who their customers are. Customers are featured prominently in all three examples. Although the mention of customers in these mission statements might appear to be insignificant at first glance, in reality the opposite is true. To set the right tone and convey the right expectations, the mention of customers in the mission statement is critical.

Guiding Principles

Guiding principles establish the framework within which a company pursues its vision and carries out its mission. Each guiding principle represents an important company value. Taken as a whole, all the guiding principles represent the company's value system—the foundation of its corporate culture. If customer service is important, all employees must see it as a company value and must help incorporate it into the company's culture. Following are examples of guiding principles one would find in the strategic plan of companies that practice ECS:

- ABC Company will uphold the highest ethical standards in the business.
- ABC Company's highest priority will be customer satisfaction.

- ABC Company will provide its customers the highest-quality products and services.
- ABC Company will continually improve its people, processes, and products.
- ABC Company will be a good corporate neighbor/citizen in the community.

All these guiding principles speak to customer service in one way or another. Two address customer issues directly; the rest do so indirectly. For example, upholding high ethical standards, continual improvement, and being a good corporate neighbor all play important roles in creating loyal customers.

Broad Strategic Goals

A company's broad strategic goals translate its mission into measurable terms. Each goal represents an actual performance target to which the organization plans to commit resources. For this reason, no strategic plan is complete without at least one customer-related strategic goal. Such a goal might read as follows:

- To consistently exceed the expectations of customers and retain them for life.

Strategic goals are broadly stated. They explain what the company hopes to accomplish, not how it will go about it. Specific actions to be undertaken to accomplish strategic goals are set forth in short-term plans such as annual work plans. These actions may change frequently depending on their effectiveness, but the strategic goal they support does not. Strategic goals, such as the one in this example, are typically worked toward but never fully accomplished. This is because no matter how successfully a company creates loyal customers or accomplishes any other strategic-level activity, it must do even better as time goes by to stay ahead of the competition.

JOB DESCRIPTIONS FOR ALL EMPLOYEES

One of the ways management communicates the message to employees that "this is important" is by putting whatever "this" is in the company's job descriptions. If management wants to emphasize the importance of customer service, customer service responsibilities should be explained in employees' job descriptions. For example, the following statement might be found in the job description of any employee in an engineering, manufacturing, or construction company:

- Employees in this position are responsible for doing everything within their range of authority to ensure that our company exceeds customer expectations every time and consistently over time.

This statement makes it clear that customer service is an important part of the employee's job. Different levels and types of employees have different ranges of authority, but within these ranges, employees are clearly expected to go the extra mile for customers every time there is an interaction. If a customer's problem falls outside of an employee's range of authority, the employee should put the customer in contact with someone who has the authority to deal with the problem. For employees who have customer-service responsibilities built into their job descriptions, "I can't help you" is never an appropriate response. Even the lowest-paid employee with the least authority can respond, "Let me introduce you to someone who can help you."

PERFORMANCE-APPRAISAL PROCESS

There is a wise saying about losing weight that goes like this: "If you want to lose weight, you have to step on a scale." This does not mean that the act of stepping on a scale will cause you to lose weight. Rather, it means that in order to lose weight, you must monitor and measure your progress. This concept applies to most performance-oriented endeavors. If you want to improve performance, you must monitor and measure it. This principle is the foundation of the performance-appraisal process.

Companies that want to be customer oriented should monitor and measure their employees' customer-service behaviors as part of the performance-appraisal process. In addition to the criteria typically found on performance-appraisal instruments (e.g., quality of work, quantity of work, job knowledge, initiative, punctuality, and so on), there should also be customer-service criteria. Figure 2.3 shows a portion of a performance-appraisal instrument containing a customer-service criterion.

The ratings assigned employees on performance appraisals such as the one in Figure 2.3 can tell an interesting story. For example, the lower two scores (3 and 6) are assigned to individuals whose attitudes toward customer service are pointedly negative. Individuals assigned these scores think customer service is the responsibility of someone else or that it is just for show. "It's not my job" summarizes the attitude of such employees toward customer service. Experience has shown over and over that employees who have this kind of attitude seldom are turned around. Even with counseling, mentoring, and training, their performance rarely reaches an acceptable level. This is because their poor performance is due more to a negative attitude than to a lack of ability. Such individuals are like the cranky librarian who thought his library would be a great place if only the pesky students would leave his books alone.

At the high end of the performance spectrum is the customer-service "champion." Individuals in this category understand that their company's success depends on customers. Consequently, they have made it their business to learn what customers expect and to go the extra mile in making sure they exceed those expectations. Customer-service champions make a point

Performance Evaluation

Employee: Division: Date:

Job Title: Department:

Rating Factors (Use any rating number from 1 to the maximum shown in each category)

Quality of Work

3	6	9	12	15	Rating
Almost always makes errors. Quality is someone else's responsibility.	Quite often makes errors. Quality is not important. It's just for show.	Makes errors, but equals job standards. Employee is quality oriented.	Makes few errors, has high accuracy. Makes quality and productivity suggestions.	Almost never makes errors. Actively participates in performance enhancement activities.	______

Quantity of Work

3	6	9	12	15	Rating
Almost never meets standards.	Quite often doesn't meet standards.	Volume of work is satisfactory.	Quite often produces more than required.	Always exceeds standards; exceptionally productive.	______

Job Knowledge

3	6	9	12	15	Rating
Inadequate.	Requires considerable assistance/ training.	Adequate grasp of essentials; some assistance required.	Knowledge thorough enough to perform without assistance.	Expert in all phases of work expected.	______

Customer-Service Performance

3	6	9	12	15	Rating
Almost never puts any effort into satisfying customers. Customer service is someone else's job.	Only occasionally puts any effort into satisfying customers. Customer service is not important. It's just for show.	Has gaps in performance, but is generally customer-oriented and does try.	Customer-oriented and can be depended on to deal positively with customers.	Customer-service champion who always goes the extra mile to satisfy customers.	______

FIGURE 2.3 Performance-appraisal form with a customer-service criterion.

Three-Step Model for Connecting Employee Rewards and Customer Service

- Decide which customer-service behaviors are expected.
- Decide how the expected behaviors will be measured.
- Train supervisors to consider the results of the measurements when recommending salary increases, incentive pay, and promotions.

FIGURE 2.4 Companies benefit from linking customer service and employee rewards.

of interacting positively with customers every time and consistently over time. Managers can use their customer-service champions to mentor those employees who are rated at the intermediate levels (9 and 12). Individuals who score in the intermediate range have the right attitude, which is the key ingredient in effective customer service, but they need to develop their interaction skills more fully. Through training, mentoring, and counseling they may become customer-service champions.

REWARD AND RECOGNITION SYSTEMS

If customer service is to be fully integrated into a company's culture, positive customer-service behaviors must be rewarded and recognized. Rewards focus on pay and promotions. Recognition consists of the many and varied ways companies publicly acknowledge the performance of selected employees.

Reward Systems

Pay and promotions are important to employees. Customer satisfaction is important to companies. There is an obvious win–win connection here. Figure 2.4 describes how companies can effectively make this connection.

Expected Customer-Service Behaviors

It is critical that employees be informed of the company's expectations relating to customer service. In fact, it is the theme of this chapter. This critical step can be accomplished as follows:

- Integrate customer service into the company's strategic plan.

- Integrate customer service into all job descriptions.
- Integrate customer service into the performance-appraisal process.
- Integrate customer service into reward and recognition systems.
- Develop a customer-service philosophy statement.
- Ensure that managers and supervisors set a positive example.

Measure Expected Behaviors

Employees' customer-service behaviors should be monitored on a daily basis by supervisors. In addition to daily observation and immediate correction as needed, formal measurements should also be used. These formal measurements include the following: (1) performance appraisals, (2) one-on-one customer feedback sessions, (3) written feedback solicitations, (4) third-party feedback sessions, and (5) real-time feedback.

- *Performance appraisals.* The most obvious form of measurement is the employee performance appraisal, in which supervisors use the knowledge gained by their daily observations to assign a numerical or qualitative score. A score that reflects either positively or negatively on an individual should be given due consideration when making recommendations for salary increases, incentive pay, and promotions.

- *One-on-one customer feedback sessions.* In these sessions, representatives of the customer are invited to meet with the the company employee responsible for their project and to give frank and open feedback. The company representative should ask such questions as: What did we do well? What could we have done better? Who was especially helpful to you? Who was not helpful, or who could have been more helpful? Feedback that reflects positively or negatively on an individual, team, or department should be given due consideration when making recommendations for salary increases, incentive pay, and promotions.

- *Written feedback solicitations.* Most people are accustomed to seeing customer-feedback cards in restaurants and hotels. Although such cards do not apply directly in engineering, manufacturing, and construction settings, the concept of soliciting written feedback from customers does. Figure 2.5 is an example of a feedback form that can be sent to the appropriate customer contact by engineering, manufacturing, and construction companies. Feedback that reflects positively or negatively on an individual, team, or department should be given due consideration when making recommendations for salary increases, incentive pay, and promotions.

- *Third-party feedback sessions.* In these sessions, representatives of the customer organization meet with a third-party consultant, who facilitates the meeting. The rationale for using a third-party facilitator is to encourage

DTS, Inc.
Customer Feedback

On the following project ______________ how well were you served? What can we do better?

Your comments:

______________ ______________

Name/Title Date

Return to:
David Stanley, CEO
DTS, Inc.
stanley@DTS.net

FIGURE 2.5 Customer feedback cards can provide important information about customer-service performance.

openness on the part of the customer representatives. Human nature is such that some people find it easier to be more frank with a third party than they do with a representative of the company. When using third-party facilitators, it is important to ensure that the desired information is solicited. Company officials can accomplish this by developing a set of questions to be asked by the facilitator and by providing the facilitator with background information on the project in question. The facilitator conducts the meeting, records the feedback, and gives the company a written report. The report is treated as

DTS, Inc.
Customer Feedback Record

Project: ______________________________

Date: ______________________________

Feedback from: ______________________________

Recorded by: ______________________________

Record customer feedback in this space:

____

____

____

____

____

____

____

____

____

FIGURE 2.6 "Real time" customer feedback should be recorded and used.

confidential. Feedback that reflects positively or negatively on an individual, team, or department should be given due consideration when making recommendations for salary increases, incentive pay, and promotions.

- *Real-time feedback*. During the course of a project or contract, company officials receive a good deal of "real-time feedback" from customers. This is feedback given during the course of a project. Real-time feedback may be the most valuable source of customer feedback. It is unfortunate, then, that this valuable feedback is often lost in the rush to solve problems and keep the project on schedule. To facilitate the collection of real-time feedback, companies can use a Customer Feedback Record such as the one shown in Figure 2.6. The form may be printed, saved electronically, or both. Whatever the case, real-time customer feedback should be recorded, tied to a given project, acted on, and used to measure employee performance. By maintaining records of real-time customer feedback tied to a given project, the company gains two valuable capabilities. First, it can analyze the records to identify trends. If negative employee behaviors of the same type are happening frequently enough to establish a trend, they could indicate a companywide or departmentwide problem. Second, it can tie negative employee behaviors to a given project. Do these negative behaviors occur more frequently in certain types of projects than in others? If this is the case, the company will be able to investigate and determine why. At the very least, the company will be able to predict, and therefore take steps to prevent, problems related to the kinds of projects in question. For example, a manufacturing firm might find by analyzing its records of real-time customer feedback that more customer problems occurred in jobs involving a certain process than in any other type of job. By investigating further, the company would be able to determine that (1) the process operators need training and (2) the process equipment needs upgrading.

Train Supervisors to Consider the Results of Measurements

It makes little sense to monitor and measure employee performance as it relates to customer service (or any other performance issue) and then to ignore the results. Yet this is precisely what happens in many companies. Because much of what is learned from the various monitoring and measuring methods should eventually be included in the performance appraisal process, the ironic gap that often exists between collecting performance data and using it to improve performance is worth examining.

Some people simply do not know how to conduct an effective performance appraisal. Others are uncomfortable measuring the performance of employees they supervise. This discomfort can be attributed to one or more factors: (1) discomfort with confrontations that might occur if an employee disagrees with a rating; (2) fear of having a negative effect on an employee's career and livelihood; and (3) fear of damaging work relations.

Performance appraisals must be objective. To ensure objective appraisals, anyone who supervisors even one employee must be given the training necessary to do the following: (1) base rating on facts; (2) avoid personality bias; (3) avoid extremes in assigning ratings; (4) avoid the halo effect; (5) avoid pecking order bias; and (6) conduct an effective face-to-face appraisal conference.

- *Base ratings on facts.* The ratings assigned in a performance appraisal instrument should be based on documented facts. This is why daily monitoring and the use of various measurement methods are so important: they document facts over time as they relate to the customer-service behaviors of employees. For example, positive or negative feedback from customers, received from a third-party facilitator's report, as real-time feedback, or in response to a solicitation for written feedback, provides supervisors with documented facts that can be used for determining ratings. If a supervisor observes during daily monitoring either positive or negative employee behavior toward a customer, the fact should be recorded immediately in a confidential log or calendar, or in some other way that can ensure confidentiality. Then, of course, action should be taken immediately either to correct the negative behavior or to acknowledge the positive behavior. The record of this event should then be referred to when conducting the next scheduled performance appraisal for the employee in question.

■ *Avoid personality bias.* Supervisors can find it difficult to be impartial when evaluating people who are different than themselves. At the same time, they can submit to the natural human tendency to rate favorably those who are like themselves. This tendency is known as "personality bias." Supervisors should keep in mind the potential for personality bias when assessing the customer-service behaviors of employees. Before assigning a rating, it is helpful to ask: "Is the rating I am about to assign influenced in either direction by personality bias?" Ratings should be based on factual information and unbiased observation.

■ *Avoid extremes in assigning ratings.* Some supervisors develop reputations for being tough evaluators, whereas others are known as soft touches. One extreme is as bad as the other. The only reputation a supervisor should have with regard to performance appraisals is a reputation for objectivity. When assigning ratings, supervisors should ask themselves if they have the documentation to back up an extreme rating. Ratings should be assigned objectively and fairly.

■ *Avoid the halo effect.* The halo effect is a phenomenon in which an employee's strong points cause the supervisor to overlook that employee's weak points. For example, an employee who is a top performer in the technical aspects of his job might be rated more highly than is deserved on the customer-service criterion. The supervisor is, in effect, putting a halo on the employee. Supervisors should be alert to avoid the halo effect every time they assign a rating on a performance appraisal form.

■ *Avoid pecking-order bias.* In any organization there is a pecking order. In every department or functional unit, some jobs are more important to the supervisor than others. Because of this, supervisors are subject to the pitfall of pecking order bias; that is, rating employees in the more important jobs higher than those in the less important ones. Performance appraisals should be ratings of employees in their individual jobs, not ratings of the relative importance of jobs. Employees in more important positions should be rewarded with higher pay, not with inflated performance appraisals.

■ *Conduct an effective appraisal conference.* The face-to-face appraisal conference between the employee and the supervisor is a critical piece of the overall performance appraisal process. In the conference, the supervisor reviews with the employee the results of the written performance appraisal. This allows the supervisor and the employee to communicate face-to-face about their perceptions, discuss those perceptions, and iron out any differences that might exist. It also gives the supervisor an opportunity to give the employee valuable feedback about her performance and to suggest ways in which the employee might improve. Finally, a face-to-face conference allows the supervisor and the employee to discuss and agree on a plan for improved performance.

Recognition Systems

Recognition is one of the strongest human motivators. People don't just want recognition, they need it. There is a sense in which recognition is the foundation of customer service. When a company goes out of its way to satisfy customers, it is recognizing the customers' importance. When customers are treated well, they know that the company understands how important its customers are. Because employees tend to treat customers the way they, the employees, are treated by management, it is critical that managers properly recognize employees.

Perhaps the best example of effective employee recognition is found in the military. The system of military commendations and decorations (medals) is based on the positive human response to recognition. No amount of pay could compel young soldiers, sailors, airmen, and Marines to perform the acts of bravery that are commonplace in the U. S. military. But the recognition of a grateful nation and grateful comrades-in-arms spurs military personnel to incredible acts of valor every time our country is involved in an armed conflict. There is a lesson here for nonmilitary organizations.

There are many ways to recognize employees who go the extra mile to satisfy customers or who perform any aspect of their job especially well. What follows are just examples.[1]

1. Write a letter to the employee's family, recounting the excellent job the employee is doing.
2. Arrange for a senior-level manager to have lunch with the employee.
3. Have the CEO of the organization call the individual or stop by personally to say, "Good job."
4. Find out what the employee's hobby is, and publicly award him a gift relating to that hobby.
5. Designate the best parking space in the lot for the "employee of the month."
6. Create a "wall of fame" to honor outstanding performance.

These examples are meant to trigger ideas; they should not be viewed as a comprehensive list. There are many other creative ways to recognize excellent performance. Every company should develop its own locally tailored recognition options. When doing so, the following rules of thumb are helpful:

1. Involve employees in identifying the forms of recognition to be used. Employees are the best judges of what will motivate them.
2. Change the list of recognition activities periodically. The same activities used over and over will lose their value.
3. Have a variety of recognition options (a "menu") for each of various levels of performance. This allows employees to select the type of recognition that appeals to them most.

CUSTOMER-SERVICE PHILOSOPHY STATEMENT

One of the most effective ways to set a customer-service-oriented tone and to establish companywide expectations is to develop, publish, and broadly disseminate a customer-service philosophy statement. Such a statement is a written summary of what the company expects of itself and commits itself to in terms of customer service. Figures 2.7 and 2.8 are examples of customer-service philosophy statements.

It is important to distribute the statement to all employees and to inform them that it represents the expectations of the CEO. It is also important to display the statement where customers can see it when they visit the company. Making the statement readily visible to customers is a key part of the accountability component of the company's customer-service program. If customers see the statement, they can call attention to it whenever an employee's behavior falls short of the expectations it declares.

In Figure 2.7, the MADAC Manufacturing & Engineering Company describes eight concepts that are to characterize its interaction with customers. These characteristics are organized into four groups, as follows:

- reliability and dependability
- empathy and caring
- responsiveness and promptness
- openness and willingness to listen

Any company that consistently lives up to the expectations set forth in MADAC's customer-service philosophy statement is likely to have excellent relations with its customers.

In Figure 2.8, the Southeastern Construction Company describes five concepts that are to characterize its interaction with customers. Attentive listening, consistency, promptness, courtesy, and dependability are all characteristics that the customers of any company would appreciate.

These two examples appear to be different at first glance, but closer examination will reveal that both companies are committing to essentially the

Customer-Service Philosophy
MADAC Manufacturing & Engineering Company

The managers and employees of MADAC are committed to providing effective customer service every time and all the time. To this end, we commit ourselves to the following principles:

➢ **Reliability and Dependability**
Our customers can rely on us to dependably provide quality services, deliver quality products, and fulfill all of our obligations with a positive attitude.

➢ **Empathy and Caring**
We care about our customers and treat them the way we want to be treated when we are customers.

➢ **Responsiveness and Promptness**
We respond promptly to our customers and are eager to assist them. This applies to in-person, written, telephone, email, and webpage inquiries.

➢ **Openness and Willingness to Listen**
We are open to input and feedback from our customers and will listen to and act on their concerns.

FIGURE 2.7 One example of a customer-service philosophy statement.

same characteristics. This is as it should be. Customers are customers, and they like to be treated well regardless of the type of company in question.

Developing Customer-Service Philosophy Statements

There are three effective ways for a company to develop a customer-service philosophy statement: (1) the company's executive management team develops the statement; (2) the company establishes a cross-functional team of employees to develop the statement, which is then approved by the executive management team; and (3) the company establishes a team consisting

Customer-Service Philosophy
Southeastern Construction Company

Southeastern Construction Company (SCC) places a high priority on customer satisfaction. SCC is committed to the following principles:

➢ **Attentive Listening**
We listen attentively to what our customers have to say, and we use it to continually improve.

➢ **Consistency**
We treat our customers with empathy and respect the first time and every time thereafter.

➢ **Promptness**
We respond to customers promptly and value their time.

➢ **Courtesy**
We treat our customers with courtesy and professionalism.

➢ **Dependability**
We make sure our customers can depend on our employees, our products, and our services.

FIGURE 2.8 A second example of a customer-service philosophy statement.

of company representatives and customers to develop the statement, which is then approved by the executive management team. Each of these approaches has its benefits and drawbacks.

1. *Executive management team develops statement.* With this approach, the company's executive managers decide what types of customer-service behaviors they expect from themselves and their employees. They then write statements describing these behaviors. Collectively, these statements represent the company's customer-service philosophy. The philosophy statement is distributed to all employees with appropriate communication about its importance, and it is prominently displayed so that employees, customers, and potential customers can see it. The advantages of having the executive

management team develop the customer-service philosophy statement are speed and authority. Executive managers can quickly draft a comprehensive statement that reflects their expectations. In addition, a statement that comes directly from the top carries inherent authority. Employees know that it's important. The downside of this approach is its inherent weakness in terms of employee buy-in. Employees can be reluctant to accept something handed down to them.

2. *Cross-functional team develops the statement.* With this approach, a team is formed with at least one representative from each division, department, and functional unit in the company. The team is then given its charter as well as appropriate direction by the CEO (timetable, how to submit drafts and to whom, and so on). The team may select its own chair, or the CEO may appoint one. The customer-service philosophy statement is developed in draft form and submitted to the CEO (or her designate) for approval. The CEO and executive managers may make additions, deletions, or corrections. Once the draft has been finalized and approved by the CEO, she distributes it to all employees with appropriate supportive communication. It is also displayed prominently so that employees, customers, and potential customers can see it. This approach has the advantage of employee buy-in and the disadvantage of slowness. Employees are more likely to accept a philosophy statement they had a hand in developing. On the other hand, forming and chartering a cross-functional team takes time, as does the step of each member of the team going back to his functional unit to solicit ideas, collect input, and solicit feedback about preliminary draft statements.

3. *Company/customer team develops the statement.* With this approach, the company's executive management team identifies company representatives and several customers to serve on the development team. Because customers are involved, one of the company representatives should be a member of the executive management team. This individual may chair the team or let the team select its own chair. Occasionally such teams select a customer as the chair. This approach to developing the customer-service philosophy statement has the advantages of double buy-in (that is, from both employees and customers) and credibility. No customer-service philosophy statement is as credible as one developed with the assistance of customers. On the other hand, this is typically the slowest approach to developing a statement. In addition to the time required to establish and charter the team, the limited availability of customers for meetings must be factored in.

SET A POSITIVE EXAMPLE

Management personnel, including first-line supervisors, set the tone and expectations relating to customer service. Integrating customer service into the company's strategic plan, job descriptions, performance-appraisal process, and reward and recognition systems is important. So is developing a customer-service philosophy statement. In fact, these steps are critical. But nothing is more important than managers setting the right example. Managers who fail to set a good example of customer service undermine all the company's other efforts to ensure customer satisfaction. Employees need to see managers upholding every aspect of the customer-service philosophy statement every time they interact with customers. Anything less than that will lead employees to consider the company's efforts nothing but talk.

Summary

1. A comprehensive strategic plan for a company has several components: a vision statement, a mission statement, guiding principles, and broad strategic goals. Customer-service expectations should be fully integrated into all of these components.

2. Customer-service expectations should be clearly explained in the job descriptions of all employees. An example of such a statement for an employee of an engineering, manufacturing, or construction company is as follows: "Employees in this position are responsible for doing everything within their range of authority to ensure that our company exceeds customer expectations every time and consistently over time."

3. Companies that expect to see improvements in the area of customer service must make it an important criterion in performance appraisals. Performance appraisal forms should contain one or more customer-service criteria. Supervisors should consider these criteria; when making recommendations for promotions, salary increases, incentive pay awards, recognition, and other types of rewards. Supervisors should be trained to base ratings on facts; to avoid personality bias, extremes in assigning ratings, the halo effect, and pecking-order bias; and to conduct an effective face-to-face appraisal conference.

4. To set the right tone for customer service, companies must ensure that customer-service behaviors are important determining factors in employee recognition and rewards. Employees intuitively know that if it pays, it must be important. The obverse is also true. If customer service does not pay, it must not be important.

5. A comprehensive customer-service philosophy statement helps set the right tone and expectations. Such a statement is a written summary of what a company expects to commit itself to in terms of customer service. A customer-service philosophy statement may be developed by the company's executive management team, a cross-functional team within the company, or a team consisting of company representatives and customers.

6. No matter what else managers do to set the tone and expectations for customer service, they must set a positive example. Failing to do so can undermine the effectiveness of all other strategies.

Key Phrases and Concepts

Avoid extremes in assigning ratings
Avoid the halo effect
Avoid pecking-order bias
Avoid personality bias
Base ratings on facts
Broad strategic goals
Conduct an effective appraisal conference
Customer-service philosophy statement
Expected customer-service behaviors
Guiding principles
Job descriptions
Measuring expected behaviors
Mission statement
One-on-one customer feedback
Performance-appraisal process
Performance appraisals
Real-time feedback
Recognition system
Reward and recognition systems
Set a positive example
Third-party feedback sessions
Vision statement
Written feedback solicitations

Review Questions

1. Explain how a company should go about integrating customer service into its strategic plan.
2. Demonstrate how a company can integrate customer-service expectations into its job descriptions.
3. How can a company integrate customer service into its performance-appraisal process? How can supervisors ensure objectivity when conducting performance appraisals?
4. How can a company integrate customer service into its reward and recognition systems?
5. Explain the various ways a company can go about developing a customer-service philosophy statement.
6. Why is it important for managers to set a good example of positive customer-service behaviors?

ECS APPLIED: SETTING THE TONE AND EXPECTATIONS AT DIVERSIFIED TECHNOLOGIES COMPANY

In the last installment of this case, David Stanley and his vice presidents at DTC decided to implement ECS companywide. Having made the decision, the company's executive managers spent a month communicating with all of DTC's employees about ECS, why it is important, and what its implementation would mean to them. In this installment, Stanley and his team begin to set the tone and expectations for the company.

"Let's begin the meeting with updates," said David Stanley. "Have you all completed your assignments?" Looking around the table, Stanley got nods from all of his vice presidents. "So job descriptions and performance-appraisal forms have been revised?" Again Stanley got affirmative nods. Tim Wang, vice president for manufacturing, added, "All employees have their new job descriptions, all supervisors have the revised performance-appraisal forms, and all employees have been shown the new performance-appraisal criteria. The human resources department did an excellent job on its assignment." "Good," responded Stanley. "Then we are making progress."

Wang continued, "The human resources director has scheduled an all-day training session for supervisors on how to conduct effective, objective performance appraisals and how to use the results when making recommendations for rewards and recognition." "Excellent. When is it?" asked Stanley. "We've already got it on your calendar," answered Wang. "We worked out the date with your secretary. It's next week, and we will all be there." "Good job Tim," replied Stanley.

After discussing several other details, Stanley said, "Well, I guess it's time for me to report on the progress I've made with my assignment." "This had better be good," said Meg Stanfield with a laugh. The vice president for engineering had wanted to handle developing the customer-service philosophy statement as part of her assignment, but Stanley had insisted on doing it himself. "Don't worry. You're going to like what I've done. I've formed a team of representatives from each of your areas and all other functional departments. But the good part is, I also have four customer representatives on the team." "You're kidding," said Conley Parrish. "Not at all. The team will have its first meeting tomorrow over lunch in the private dining room at Mancotti's Restaurant downtown, and Meg, you are going to chair the team." "I do like it," said Meg Stanfield, obviously pleased. Stanley asked Stanfield to make sure the customer representatives participate in a frank and open way.

"Now I want to talk about what might be the most important aspect of this whole effort," said Stanley. "I am concerned about all of us setting the right example. This is going to be critical. Our employees will take their cues from us and other management personnel. We have got to get this right every time we interact with customers, or all of the other things we are doing will go up in smoke." Stanley went on to tell the vice presidents about several times he had "lost his patience" with customers in front of employees. He also reminded them of times when they had done the same thing. Then he asked each of the vice presidents to identify the management and supervisory personnel in their respective divisions who might be problematic in setting a positive example. Several were identified in each division.

"What should we do about them?" asked Stanley. "I think we have to do at least a couple things," commented Tim Wang. "Each of us needs to have one-on-one conversations with the potential problem people in our divisions to let them know what we expect. In addition, we need to monitor them closely and evaluate their performance in terms of customer service. They need to know that being a good engineer or a good technician is no longer enough—customer-service performance is also important, and it's going to count. They won't know these things until we show them, and I guarantee they will test us." Stanley nodded indicating agreement and then asked his vice presidents to make setting the right example a high priority for all levels of management. "Let's report on this issue at all future meetings."

DISCUSSION CASES

The following cases provide examples of how the various concepts presented in this chapter might play out in actual companies. The cases are provided to prompt discussion, give the reader a feel for the types of problems confronted in the workplace, and reinforce the ECS concept in question.

CASE 2.1 "It's Not My Job" Is No Longer an Acceptable Response

Parker Engineering Company (PEC) is the largest engineering firm in a small community. Consequently, when James Parker, the company's founder and CEO, goes out to lunch or to the grocery store he is likely to bump into a customer. When this happens, it is not uncommon for his customers to give him feedback about their dealings with PEC. Several months earlier, Parker had slipped into a local diner for lunch and found himself sitting across the aisle from a customer, Marcus Ortega, a building contractor. "How's it going Marcus? Are you staying on schedule on the new shopping center project?" Parker, who was in a hurry and didn't want to get bogged down in conversation, thought this was a safe question, because Ortega is known by all in the community as an excellent project manager. Consequently, he was surprised when Ortega said, "I might be able to stay on schedule if I could get some help out of your people. When I call your civil engineering section, all I get is the runaround. What's going on with your people anyway, James?"

Caught off guard, Parker could only mumble an apology, gulp down the rest of his lunch, and dash off to a meeting. But the brief encounter with Ortega gnawed at him for the rest of the day. That night Parker called Ortega and asked if they could meet for breakfast the next day. They did, and Ortega let Parker know that he was really unhappy with PEC. "It's like you've gotten too big to listen. You're so busy chasing new customers, you've forgotten about your old ones." Parker didn't argue. Instead, he asked Ortega to give him a week and then call the company again.

Parker used the week well. First thing Monday morning, he called all of his employees together and put a stack of papers on the conference table and said, "These are your job descriptions. I spent several hours last night going through them. There is not even one word about customers or customer service in these job descriptions. That's my fault, and it is going to change as of today. As soon as we finish this meeting, I'm asking my secretary to add the following statement to every job description in this company, beginning with mine." Parker read the statement he had prepared and asked for comments. There were several, and Parker revised the statement accordingly. By the end of the day, every employee at Parker Engineering Company had a new job description with a statement that read as follows: "Customer satisfaction is our company's highest priority and a fundamental responsibility of individuals in this position." Parker then told the employees about his encounter with Marcus Ortega, but he withheld Ortega's name. Parker knew if he revealed his name, Ortega would get better treatment in the future, but Parker wanted more than that. He wanted every customer to get better treatment. "From now on, I want every call we get from a customer to be treated as a personal challenge. Every one of us is to take responsibility for ensuring that the customer is well served no matter who answers the phone and no matter who the official PEC contact is."

Marcus Ortega's next telephone call to PEC turned out to be a vastly different experience than those of preceding months. Pleased with the change, he invited James Parker to join him for a follow-up breakfast during which he said: "You asked me to give you a week and then call back. I did, and the change was noticeable. What did you say to your people?" "Not much," laughed Parker. "I just explained the concept of customer service."

Discussion Questions

1. Have you ever dealt with someone in a company who acted as if your problem "was not my job"? Explain.
2. What would this company have to do to win your confidence?

CASE 2.2 Good Engineer, Bad Attitude

Janice Baker could not remember ever being so angry. The promotion she wanted had gone to another engineer, Martha Andrews. "Andrews is good," mumbled Baker under her breath, "but I'm better, and everyone in this company knows it." This was more than just anger speaking. Baker *was* good. In fact, when it came to design, Janice Baker was undeniably her company's best engineer. Martha Andrews was also an excellent designer, but even she would acknowledge Baker's slight edge when it came to design.

In addition to being a good designer, Janice Baker worked well with her colleagues and was highly respected in her field. She had written several articles for engineering journals and presented papers at prestigious national conferences. Baker liked engineering and she liked engineers. She was known for putting in extra time at night and on weekends to help her colleagues keep their projects on schedule. Janice Baker had a loyal following among her fellow engineers. Consequently, it struck her as odd that there was no groundswell of protest when the company selected Martha Andrews over her for promotion to engineering manager for the mechanical engineering department. Her colleagues seemed to have accepted Andrews as the logical choice for the promotion. It was if they knew Baker would not get the job.

Baker was so shocked and disappointed that for days she couldn't keep her mind on her work. Out of frustration, Baker took a week of vacation to see if she could settle down and get things into better perspective. During this week she had lunch with another engineer from the company who she respected and trusted. John Macon was a rising star in the company who Baker had mentored and helped through some difficult projects. Baker knew Macon would be frank in helping her sort through the confusion she felt over missing out on the expected promotion.

Janice Baker really had only one question for Macon: "Why?" Expecting the question, Macon got right to the point. "It's the customer-service issue, Janice. Nobody in the company is better than you at engineering, but nobody is worse when it comes to customer service. When it comes to dealing with customers, your reputation can be summed up like this: good engineer—bad attitude." "I don't have a bad attitude!" Janice Baker almost shouted. "Not with your colleagues, Janice. Just with customers." "I don't like dealing with customers," acknowledged Baker. "Half the time they just don't get it. They can be illogical, ill informed, and pushy. I'd rather deal with engineers. They make sense. Let the guys in marketing deal with customers."

Macon went on to explain that everyone in the company knew of Baker's feelings toward customers. "That's why you are rarely invited to participate in meetings with customers," said Macon. "That doesn't bother me," shrugged Baker. "While the others are schmoozing with customers, I'm back in the office keeping their projects on schedule." "That's right, and everybody respects you for it," smiled Macon. "You're a behind-the-scenes engineer, Janice. In fact, you're the best there is in that setting. But to be the manager of an engineering department, you've got to be good at meeting with customers. Coordinating with customers is a critical part of an engineering manager's job at our company."

Baker was quiet for a long time as she stared across the restaurant at nothing in particular. Macon let her think without interrupting. Finally, she turned to him and said, "John, I guess the company made the right choice. I don't like dealing with customers, and I never will. Martha Andrews will do a good job as engineering manager." Baker was right on both counts. Andrews turned out to be even better as an engineering manager than she had been as a project engineer, and Baker continued to be the company's top designer. She never did become comfortable with customers, but the company structured her job to minimize Baker's contact with customers while maximizing her contact with fellow engineers.

Discussion Questions

1. Have you ever worked with someone who was good at her job but had a bad attitude? Explain.
2. Did this person's attitude affect you, customers, or fellow employees? If so, how?

CASE 2.3 You Remembered the Stick but Forgot the Carrot

Mark Cisco is plant manager and senior engineer at the Oklahoma plant of Man-Tech, Inc. The company has 10 plants located in 10 different states. In total, Man-Tech's plants employ from 900 to 1,500 people and manufacture voting machines for local governments. He and his fellow plant managers from Man-Tech's other plants meet once a month to discuss problems and share best practices. Typically, the discussion covers engineering and manufacturing topics, but recently Cisco introduced the issue of customer service, and he was surprised to find a high level of interest in the topic among his colleagues.

The consensus of the group was that customer service is (1) critically important in a competitive marketplace, (2) given insufficient attention in most colleges of engineering, and (3) difficult to do well. Cisco described for his colleagues the various components of a customer-service program he had implemented at his plant. In an attempt to improve customer service, Cisco had rewritten his plant's strategic plan (Man-Tech's plants operate autonomously within prescribed marketing territories) to include a commitment to customer satisfaction, added customer service responsibilities to the job descriptions of all employees, and integrated customer service into the plant's performance-appraisal process.

"How's the program working?" asked Ed Visey, plant manager of Man-Tech's Ohio branch. Visey, it turned out, had implemented a similar program at his plant. "Not as well as I'd like," responded Cisco. "The program has helped, but not enough. On a scale of 1 to 10, my plant was probably a 1 before we implemented this program. Now we are probably a 5. If I'm going to stay ahead of the competition in my state, we have to be at least an 8 or a 9." "I know what you mean," said the manager of Man-Tech's Oklahoma plant. "Ever since the debacle in Florida during the Bush/Gore presidential election, it seems that every manufacturer in the world has decided to get in the game. Everybody knew the problem was wider than just Florida. Other manufacturers sensed an opportunity, and now it seems that everybody is trying to make voting machines." "I know just what you mean," said Cisco. "I used to bid against one or two companies when some county needed new voting machines. Since the Bush/Gore election, every bid put on the street has at least 10 companies chasing it, not counting Man-Tech. To make matters even worse, county elections supervisors don't want to be caught off guard, like some of their colleagues in Florida were, and be embarrassed by all the negative press attention. They have become tough customers with high expectations."

"What about the reward and recognition components of your program?" asked Visey. "What do you mean?" queried another participant. "Well, the strategic plan and job descriptions show what's expected, and that's good. The performance appraisal puts some teeth into the expectations by measuring actual performance against the expectations. This is good too. But all performance-enhancement efforts are based on the old principle of the carrot and the stick. Unless you reward and recognize employees for meeting or exceeding expectations, you've left off half of the equation. In other words, you are applying the stick, but forgetting the carrot."

Participants discussed the issue of rewards and recognition for a while, and Cisco had to admit that at his plant, positive customer-service behaviors really had no affect on raises, promotions, incentive pay, and recognition. "Customer-service behaviors are covered during the performance-appraisal process, but I don't think any of our supervisors really give these ratings much consideration when they make recommendations for rewards and recognition. Our supervisors are engineers and manufacturing technicians. They just don't seem to put a high priority on customer-service performance." "There's your problem," Visey said. "I discovered the same thing the hard way while implementing my program. My supervisors were not attuned to factoring in customer service ratings when making recommendations for raises, promotions, awards, and so on. Even though we had changed our performance-appraisal forms to include customer-service criteria, the supervisors still made no connection between the ratings on these criteria and rewards and recognition. Their focus by both training and preference is design, processes, and technical issues. I found out that people who have spent their working lives in engineering and manufacturing just don't think about customer service when they want to promote an employee or give someone a raise. We made no more progress than Mark has until we did two things. First, I met with all supervisors and let them know that I expected customer-service ratings to be given equal consideration with other criteria when recommending rewards and recognition. Second, I contracted with a consultant to provide a two-day seminar on conducting effective performance appraisals and using the results of the appraisals to improve performance. I attended the seminar myself and required all supervisors to attend—no excuses allowed. On Mark's scale of 1 to 10, my plant is now probably an 8. With a little more practice and a lot more attention from management, we'll be a 9 before long."

While Visey talked further about his program, several of his colleagues, including Mark Cisco, took notes. After the meeting, was adjourned, each of Man-Tech's plant managers returned home with a briefcase full of good ideas. Mark Cisco was on the telephone calling Visey's consultant even before his taxi made it out of the parking lot.

Discussion Questions

1. Have you ever worked for a supervisor who was quick to point out the negative, but seldom acknowledged the positive in your work? Explain.
2. Was this supervisor effective in improving your performance? Why or why not?

CASE 2.4 Let's Involve Not Just Customers, but Also Lost Customers

Angie Parks put in more than 20 years in engineering and marketing before becoming CEO of EnCon Tech, Inc. Her years in marketing had brought her in touch with customers and potential customers nearly every day. Consequently, when she decided that EnCon needed a customer-service philosophy statement, Parks was determined to involve customers in the development process. EnCon, which does manufacturing and construction and has an engineering department supporting both, has five vice presidents (engineering, manufacturing, construction, marketing, and finance/administration). Along with Parks, the five vice presidents make up EnCon's executive management team.

When Parks first broached the issue of developing a customer-service philosophy statement, EnCon's other executive managers said, "Fine, let's develop it." Parks, however, wanted to involve not just representatives from all of the company's functional units, but customers too. And not just any customers. She wanted to ask representatives from two customers EnCon had recently lost. In fact, it was the loss of these two customers—one from the manufacturing division and one from the construction division—that had convinced Parks that EnCon needed a customer-service philosophy statement. Both companies had been long-term customers of EnCon, and the CEOs had told Parks they were leaving because of poor service. Consequently, Parks was adamant that customers be involved. In addition, Parks told the vice presidents that she would chair the team and that she wanted each of them to name a representative to it. The vice presidents, possibly concerned about what the customer representatives on the team might say, made a case for serving themselves. But Parks demurred, saying the vice presidents instead would serve as the "reality check" for what she and the team developed.

It took several days and numerous telephone calls for Parks to convince EnCon's lost customers to participate, but she persevered and finally won out. At the team's first meeting, which she held away from EnCon's office at a neutral location, Parks asked the customer representatives to do two things. First, she wanted them to make a list of the key words and phrases that came to their minds when they considered how they like to be treated as customers. Then she asked them to circle the ones that were most often missing over the years when they dealt with EnCon. As for EnCon's representatives, she asked them to make a list of key words and phrases that described how they like to be treated by the company's suppliers, and then to circle those that were most often missing when interacting with suppliers. When the two groups compared lists, most participants were surprised at the extent to which they were similar.

Common to both lists were the following concepts:

- Not willing to listen
- Do not seem to care about my problems
- Not prompt in returning telephone calls, emails, and so on
- Not respectful of my time
- More concerned with selling me what you have than giving me what I need
- Apparent lack of coordination between marketing and engineering departments and between engineering and production personnel
- Negative attitudes of company representatives
- Poor telephone skills (personnel seem either bored or overly rushed)
- Missed deadlines
- Poor follow-through
- Failure to keep promises

It took Parks three months to form the development team, collect the team's input, lead participants in developing a draft statement, run the draft statement by EnCon's vice presidents, and produce a final product. But when the statement was completed, it was a worthwhile document. In fact, it was so good that Parks decided to host a ribbon-cutting ceremony during which she would unveil the new customer-service philosophy statement. She invited not just all of EnCon's employees, but also the CEOs of EnCon's most frequent repeat customers and the CEOs of the two companies EnCon had recently lost as customers. After the ceremony, the CEO of one of the lost customers approached Parks and said, "Angie, if you live up to what you say in that statement, my company will be back as a customer."

Discussion Questions

1. Have you ever had a company try to win you back as a customer after it had lost your business? Explain.
2. If so, what did the company do to try to win you back?

CASE 2.5 A Tale of Two Managers

Peter Morgan, CEO of Southern Engineering Company (SEC), decided that poor customer service was causing his company to lose business. Mark Zaffman, CEO of Zaffman Construction, Inc. (ZCI) came to the same conclusion at about the same time. Morgan and Zaffman are best friends and former college roommates. Consequently, they spend a lot of time together talking shop. Both decided to implement comprehensive programs to improve customer service, and both did. Morgan experienced encouraging results. His company not only stopped losing customers, it won a few former customers back; but Zaffman's company continued, in his words, "to bleed." Their programs were almost identical. So what was the problem with the program at ZCI? A brief profile of the two CEOs tells the story.

Morgan has always been comfortable with people. In college, Morgan involved himself in university life and was an officer in the school's engineering club. Zaffman, on the other hand, has always been uncomfortable around people. In college, he spent all his time in the library and engineering laboratories. He likes dealing with plans, equipment, and materials, but not with people. Morgan's problem at SEC had been that he failed to develop in employees his ability to work with people. Zaffman's problem at ZCI had been that he did pass his ability along—and it was negative.

Once Morgan got his program in place and communicated the importance of customer satisfaction to his employees, his example was a powerful catalyst for immediate improvements. Zaffman, on the other hand, told his employees, "I don't like this customer-service nonsense any more than you do, but we've got to do it." In the weeks that followed, employees heard Zaffman complain constantly about customers, and they even saw him approve work that clearly did not meet specifications. He often said, "We'll worry about it later." Like Morgan's, Zaffman's example was a powerful catalyst, but not for improvements. In fact, after Zaffman's half-hearted, insincere effort at improvement, customer service worsened at ZCI.

Discussion Questions

1. Have you ever worked for a supervisor who obviously disagreed with company policy and only reluctantly went along?
2. How effective was that supervisor in carrying out policy?

Endnote

[1]B. Nelson, "Secrets of Successful Employee Recognition," *Quality Digest* 16(8) (August 1996): 29.

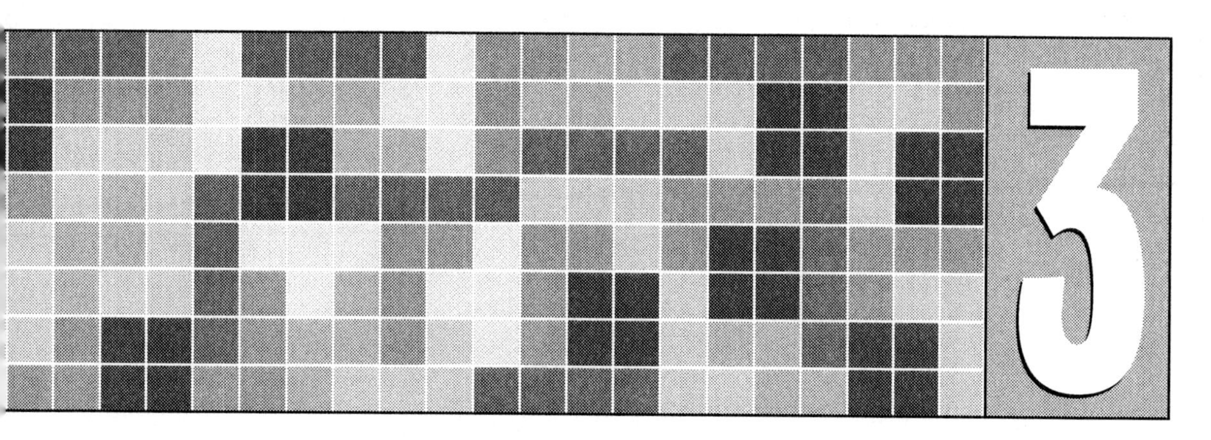

Identify What Your Customers Want

Corporate decision makers who think they know more about what their customers want than do the customers themselves are leading their companies down a one-way street to disaster.

GOALS

- Understand the relationship among Steps 3, 4, and 5 of the ECS implementation model.
- Establish an overview of customer needs.
- Learn the benefit of conducting internal reviews.
- Understand the value of assertive listening.
- Understand the role of visiting teams.
- Understand the role of the following in achieving effective customer service:

Self-evaluation	Written feedback
Customer interviews	Verbal feedback
Focus groups	Questionnaires

Customers become satisfied when their needs are exceeded by the companies they choose to work with. To exceed its customers' needs, a company must know what those needs are. Corporate decision makers who think they know more about their customers' needs than do the customers themselves are leading their companies down a one-way street to disaster.

Most people can give their own favorite examples of poor customer service. One of the authors once stood in an increasingly long line in a grocery store while a young and oblivious cashier had a ten-minute cell phone conversation with her boyfriend. The more impatient the people in line became, the more intransigent she became. When she finally hung up, she looked at her watch and said, "Oh, it's time for my break." Then she turned around and walked off. "War stories" about poor customer service typically involve poor-quality products, rude or indifferent service, bad attitudes, long waits, lack of responsiveness, poor telephone skills, insufficient follow-up, employees who lack sufficient product knowledge, and poor-quality workmanship (both in original products and in repairs or revisions).

Although these customer-service problems are common and predictable, they are not specific. Companies trying to improve customer service need to be cognizant of and concerned about these predictable problems, and they should take steps to prevent them. But dealing with these common predictable problems alone is not enough. Companies must go beyond that to focus on their customers' specific issues. This chapter explains how engineering, manufacturing, and construction companies can identify the specific needs of their customers as well as problems relating to those needs.

THE RELATIONSHIP AMONG STEPS 3, 4, AND 5

It is important that readers understand the relationship among Steps 3, 4, and 5 of the ECS implementation model. In the step covered in this chapter (Step 3), companies identify the product attributes and service characteristics that are most important to customers. In Step 4, covered in chapter 4, companies benchmark the processes that produce or affect those attributes and characteristics. Then in Step 5 (Chapter 5), companies compare the performance of the processes in question against the benchmarks established in Step 4. These three steps, taken together, enable a company to know specifically how to satisfy customers better than any other competitor can.

OVERVIEW OF CUSTOMER NEEDS

Customers want many things from the companies they work with. In the language of customer service, these "wants" are often referred to as "needs." Most customers need quality products, quality service, and friendly interaction with knowledgeable people who care about them. Think about it. Aren't these the characteristics you look for as a customer? On the surface, these needs seem fairly simple. Unfortunately, few things involving people are ever simple. What complicates customer needs are the customers themselves. Of course customers want quality products, but how do they define product quality? Of course they want quality service, but how do they define service quality? Finally, of course customers want to be treated well, but what exactly do they mean by "well"? These are questions only the customers themselves can answer. Consequently, to get the right answers, companies must query their customers directly. Asking customers what they want is the subject of this chapter, but before getting into methodologies, we must first look more closely at the subject of customer needs.

Before asking customers what they want, we need to make sure we fully understand the basics about customer needs. Almost all customer needs fall into one of the following categories:

- Policy/system/process/procedure needs
- People needs
- Value needs

Policy/System/Process/Procedure Needs

Policies, systems, processes, and procedures control how companies interact with, and deliver products and services to, customers. If you have ever had a problem with a company's hours of operation, you have experienced a policy problem. You needed the company to be open for business during certain hours, and its policy was to be open at other hours. If you have ever had difficulty getting through a company's computerized telephone system to the person you needed, you have experienced a system problem. If you have ever had to stand in line for a long time to conduct your business with a company, you have experienced a process problem. If you have ever been told that you filled out the wrong form and now must complete another form, you have experienced a procedures problem.

These are just snapshot examples of common policy, system, process, and procedure problems. Other common problems experienced by customers in these areas are the following:

- Physical issues (inconvenient location, insufficient parking accommodations, difficult ingress or egress, confusing building and grounds layout, inadequate public waiting areas, poorly maintained restrooms, uncomfortable meeting rooms, etc.)
- Inadequate communications systems (telephone, webpages, email, etc.)
- Inadequate staffing to meet customer needs
- Insufficient or nonexistent employee training

- Inconvenient or inadequate transaction or recordkeeping systems
- Outdated or inadequate technology (computer systems, CAD systems, manufacturing equipment, construction equipment, etc.)
- Inadequate warranty/guarantee policies
- Poor accounts payable or receivable policies or systems
- Inconvenient delivery/shipping services
- High-pressure marketing and sales policies and procedures
- Questionable contract policies
- Poor or nonexistent customer-interaction policies

Customers have needs and can experience problems in all of these areas. Consequently, companies trying to improve customer satisfaction should periodically examine their policies, systems, processes, and procedures. The pertinent question to ask is: How do these things affect our customers? No matter how well policies, systems, processes, and procedures are received internally, they are counterproductive unless they are well received by customers.

People Needs

Customers want to deal with employees who are courteous, respectful, responsive, knowledgeable, and caring. When employees fail to display these traits, problems are sure to follow. People problems are just as common in the engineering, manufacturing, and construction industries as they are in the retail and hospitality industries. People problems are often caused by poor communication, bad attitudes, insufficient knowledge, or a combination of these.

There is an important point that needs to be made here. At one time companies, in addressing customer-service issues, adopted slogans like: "The customer is always right." The problem with that type of slogan is that it is inherently false, and employees know it. Anyone who has ever worked with customers knows the customer isn't always right. In fact, the customer is often wrong. A better slogan, if a company feels it must have a slogan, would be: "Customers may not always be right, but they should always be treated right." Here are some all-too-common ways that customers are treated poorly:

- Employees display negative attitudes.
- Employees display negative body language, apathy, or rudeness; greet customers in an unfriendly manner; or fail to even acknowledge the customer.
- Employees have insufficient knowledge to be helpful.
- Employees give customers inaccurate information.

- Employees allow interruptions and distractions while working with a customer.
- Employees fail to listen to customers, or don't listen well.
- Employees appear rushed or distracted.
- Employees are less than professional in their appearance.

Customers are people. They want to be appreciated for bringing their business to your company. They also want to be able to trust that every time they deal with your company, they will be treated right—not an unreasonable expectation.

Value Needs

The value equation set forth in Chapter 1 describes how customers define value. Recall that the value of a product or service is a combination of the customer's perceptions of the following factors:

- Product quality
- Service quality
- People quality
- Image quality
- Selling-price quality
- Overall-cost quality

Customers who deal with engineering, manufacturing, and construction companies are sophisticated enough to consider all of these factors when doing business. Products must have the attributes customers want, and these attributes must meet or exceed customers' expectations. Services must be provided promptly, courteously, efficiently, and effectively. The company's people must display the positive characteristics customers expect. A positive image must be established, maintained, and carefully guarded. The selling price of products and services must be reasonable and competitive. Finally, the overall cost, including maintenance, warranty, upgrading, and after-purchase service, must be made readily available to customers.

INTERNAL REVIEWS

The first step in identifying what customers want can be found in the sage advice of the old monk: *Look within yourself.* In every company there are several potentially productive sources of customer information, but the information may not be readily apparent. Like ore, it may need to be mined from

Internal Sources of Customer Information

- **Warranty Records**
  Product or workmanship discrepancies or failures
- **Customer-Service Records**
  Most commonly heard complaints and comments
- **Front-Line Personnel**
  What customers say about you
- **Professional Involvement**
  Customer-related problems your colleagues in the field are experiencing

FIGURE 3.1 Companies should look within for customer information.

these sources: warranty records, customer-service records, front-line personnel, and professional involvement (see Figure 3.1).

Warranty Records

Companies that guarantee their products, services, and work have a built-in source of customer information. When a product is returned for repair or replacement, a record is made. When a crew must be dispatched to correct warranted work that did not hold up for the required period of time or to repair equipment that does not function properly, a record is made. These records, if thoroughly analyzed, contain a wealth of information on customer needs and attribute preferences.

"Attributes" are features of a product or characteristics of a service about which customers have definite preferences. For example, most automobile buyers think reliability is an important attribute. They want to know that every time they put their key in the ignition and turn it, the car will start promptly. A car that is hard to start is not reliable and, if under warranty, will surely cause complaints and produce repair work. In such cases, records are kept and can be examined to identify frequent problems or problem trends.

As another example, customers who deal with engineering firms expect to receive accurate, well-prepared drawings. Accuracy is an attribute expected of engineering plans, calculations, bills of material, and so on. Inaccuracies in any of these products lead to problems that, in turn, lead to unplanned work for which there will be records.

Using warranty records to identify product attributes and service characteristics that fail to meet customer expectations would seem to be an obvious strategy. However, there are many companies that simply file warranty records away and never look at them again. One of the authors was once asked by a manufacturer of kitchen appliances to help determine why sales of its best blender had plummeted in recent months. The company's plan was to have the author design a customer questionnaire and conduct a survey. As a first step, the author asked the company's CEO to have all warranty records for the past 12 months pulled for the blender in question. The records showed that more than 90 percent of the complaints had to do with stripped gears in the main mechanism of the blender. The author didn't need to proceed to a questionnaire or survey.

To save on material and manufacturing costs, the company had changed from metal to plastic gears. The plastic used to manufacture the gears was, according to its manufacturer's claims, supposed to "almost equal" the properties and performance of metal. It didn't. Here was a company that had warranty records readily available, but no one was assigned to study them—a surprisingly common oversight. Companies that fail to regularly analyze their warranty records ignore a potential treasure chest of customer information.

Customer-Service Records

Closely related to warranty records are customer-service records. Companies that provide customers a hotline to call or an email address to contact when they have problems have an excellent source of customer information to draw upon. Customer-service calls, unlike warranty complaints, are not necessarily about product failures or workmanship discrepancies. Often they go like this: "I just purchased your product, and I cannot get it to work. What am I doing wrong?" Software manufacturers frequently receive this kind of call. In such cases, the product will probably operate as designed, but only if customers know how to do their part. The problem could be insufficient or poor customer training (see Chapter 5), a poorly written users manual, or poorly designed operating procedures. Even the best-designed product is faulty if customers cannot figure out how to use it.

A washing machine manufacturer found this out the hard way when its engineers designed an ultramodern digital keypad to replace the old dials and knobs previously used to operate its models. The company's engineers loved the digital keypad. To them, it was convenient and easy to use. Unfortunately, few of the washing machines used in America's homes are operated by engineers. The manufacturer made the mistake of assuming that customers would like what its engineers like. Consequently, the new-model washing machines were manufactured and released to retail showrooms without so much as even a cursory market test. It was soon apparent that customers were rejecting the new models because of the digital keypad. Customer feedback was clear: "Give us back our knobs and dials!"

Market forces gave the manufacturer no choice but to undertake a costly retrofitting program in which the digital keypad was replaced with knobs and dials. The retrofitting program saved the washing machine manufacturer from extinction, but at an incredible cost to the company in terms of direct dollars, stock value, and damage to corporate image—damage the company's competitors were only too happy to exploit.

Frontline Personnel

Frontline personnel are those employees who have the most contact with customers. They typically include receptionists, clerical personnel, secretaries, sales personnel, marketing representatives, and after-purchase service providers. Personnel in these categories have access to a wealth of customer information, but too few companies regard them as legitimate sources of customer information. We use the term "mining customer information" to describe the process of collecting such information from frontline personnel. This is because, like mining, the process involves digging out the information. It can take a concerted effort. Collecting customer information from frontline personnel is no passive undertaking. It means engaging frontline personnel in periodic data-collection efforts (for instance, internal focus groups, one-on-one conferences, internal surveys, and so on). The process should be regular and systematic.

The importance of collecting customer information from frontline personnel cannot be overstated. A comparison of two competing companies makes the point. Both companies design, build, deliver, and set up portable metal buildings. The buildings are constructed of wooden frames covered by an aluminum skin. Both companies contract out the delivery and setup of their buildings. Setup involves placing the building on the buyer's property, leveling the building, and making sure that windows and doors work properly.

Company A's engineering personnel do not communicate with the delivery/setup contractor unless a window or door has been damaged in shipping and must be replaced. Company B, on the other hand, has a face-to-face meeting with its delivery/setup contractor immediately after every project. The purpose of the meeting is to record and discuss any feedback the customer may have given or any problem the contractor encountered. These debriefings provide valuable insight into customer needs and preferences. As a result of the the information collected from the debriefings, Company B has made numerous improvements to its buildings and to its delivery and setup procedures. For example, after several customers complained about having to wire their new building for electricity, the company began offering built-in wiring as an option. When numerous customers commented that they planned to use their portable building for temporary office space, the company began offering models with electrical wiring, paneling, insulation, and floor covering options (carpet, linoleum, finished wood, and so on). The various options added in response to customer feedback are popular with customers. In fact, the wired, paneled, and insulated models are now the company's best sellers. Company B's annual sales are more than four times those of Company A.

Professional Involvement

Professional-level personnel in engineering, manufacturing, and construction can gain valuable customer information through involvement in professional organizations. Professional involvement includes such activities as reading professional journals, attending conferences, and participating in meetings with colleagues in the field. It is the nature of professional journals, conferences, and meetings to deal with leading-edge issues and information. If colleagues have discovered new or neglected customer needs or new attributes that appeal to customers, this information often finds its way into professional journals and onto the agendas of professional meetings.

Professional involvement is also a good way to obtain information for establishing performance benchmarks (see Chapter 4). If an engineering, manufacturing, or construction company is to be competitive, it must perform at a competitive level. But what is a competitive level of performance? One of the best ways to address this question is using information gained through professional involvement. Professional journals typically contain articles about performance levels and how to improve them. For example, most journals on quality, engineering, and manufacturing contain one or more articles on "six-sigma" performance. Companies that must compete in pursuing six-sigma performance will find it necessary to use this as their performance benchmark. If employees of an American automobile manufacturer read in a professional journal that a Japanese manufacturer has reduced its production-line time on a competing model to just 15 hours, the American manufacturer will be forced to use 15 hours as a performance benchmark.

ASSERTIVE LISTENING

In ECS Case 3.2 at the end of this chapter, "The CEO's Secret Weapon," the receptionist, Meg Parsons, is an assertive listener. This is why she is so valuable to her CEO. Companies trying to thrive in a competitive marketplace

Checklist for ASSERTIVE LISTENING

- ✓ Maintain an affirming attitude.
- ✓ Block out distractions.
- ✓ Maintain eye contact.
- ✓ Do not react or deny.
- ✓ Ignore attitude, concentrate on content.
- ✓ Ask questions to clarify when necessary, but do not interrupt.
- ✓ Paraphrase and repeat what has been said.

FIGURE 3.2 Assertive listening takes practice.

need to train all their employees and management personnel to become assertive listeners. Figure 3.2 is a checklist of strategies for assertive listening.

As can be seen from the listening strategies in Figure 3.2, assertive listening, as the name implies, is no passive undertaking. Quite the contrary. Assertive listening requires concentration and effort.

The first listening strategy in Figure 3.2 is to adopt an affirming attitude. An affirming attitude says nonverbally to the speaker, "I am listening, I am interested in what you have to say, and your feedback is important to me," through facial expression, posture, and eye contact. Such an attitude is the opposite of the "I-don't-want-to-hear-it" attitude that is so common. Remember, the purpose of adopting an affirming attitude is to encourage customers to verbalize what is on their mind rather than to spread their discontent among other customers and potential customers.

Another assertive listening strategy is to block out distractions. Few things are more frustrating for a customer than to be constantly interrupted while trying to make a complaint or tell someone about a problem. And you will learn very little if every time a customer tries to explain a problem, she is interrupted by a telephone, beeper, unannounced visitors, or other common and predictable distractions. All distractions should be eliminated when you are listening to customers. Secretaries can hold telephone calls and detain people who drop in. Cellular telephones and beepers can be put in silent mode. If your desk is covered with distracting clutter, you can find another place to talk or put two chairs in front of your desk so that the clutter will be behind you.

The most successful card players learn to maintain what is commonly known as a *poker face*. This means they avoid letting their facial expressions or other reactions reveal what they are thinking or feeling—in this way, they avoid "giving away their hand," so to speak. The ability to maintain a good poker face is a useful skill to have when dealing with customers. It takes practice to maintain a straight face when dealing with a customer whose anger or frustration causes an internal reaction on your part, but the practice is well worth the payoff. Reacting to what a customer says or, worse yet, denying or refuting it, will only serve to make matters worse.

Once the customer begins talking, look him directly in the eye and concentrate on what he is saying. Ignore how the message is presented; rather, concentrate on content. If you get caught up in a customer's confusion, frustration, or presentation, you may miss the actual message. If customers are upset, it is important to let them vent without interrupting. Venting usually dissipates anger and frustration, a necessary first step in communicating about a problem. Anger and problem solving are antithetical. Once the customer has settled down and has related his problem, ask clarifying questions if any part of the message is unclear.

Once you have heard the message and clarified it, repeat the message back to the customer, paraphrasing their complaint. This will show the customer you have listened to and heard the message, and it gives him an opportunity to correct your understanding or fill in anything you may have missed. This step is important because you don't want to waste time and effort looking into the wrong problem or taking action based on a misunderstanding of what was said.

Customer complaints are important and should be dealt with promptly. But this is just the beginning of what should be done with customer feedback. The next step is to determine if the customer's problem is evidence of a larger problem. For example, consider the situation in which a customer complains about inaccuracies in a set of engineering plans. The first step is to get the plans corrected, preferably before the product or structure in question is improperly manufactured or built. The next step is to ask questions such as the following:

- Were these inaccuracies the result of minor oversights, or are they evidence of a training need for the CAD technicians?
- Are these inaccuracies evidence of a need for more effective "checking" procedures?
- Are these inaccuracies the result of too few personnel trying to do too much work in too little time?

Solve the customer's problem or tend to his complaint, but don't stop there. Train personnel to ask themselves, "What does this complaint really mean?" Looking beyond problems to their causes is the only way to make real and meaningful improvements (see Chapter 6).

VISITING TEAMS

Visiting teams can provide valuable information about customer needs. A "visiting team" is a group of representatives from one company that visits other companies to observe best practices. Visits to companies that are particularly effective in the area of customer service can be an effective strategy. Much can be learned by observing others who do something well and by discussing your observations with them. Of course, there is an obvious problem with this strategy: What company wants a competitor snooping around its premises and learning from its strengths?

There are several ways of dealing with this shortcoming. The first is to arrange visits to similar companies in other regions—companies far enough away that they are not likely to be competitors. Another approach is to visit companies in other lines of business. A manufacturing firm can learn much from a construction firm and vice versa, for example, or a mechanical engineering firm can learn from observing a civil engineering firm. In these examples, the customer service processes will be similar enough to be adaptable. A final approach is to promote relationship building among top executives. Executives and other professional personnel in engineering, manufacturing, and construction companies can get to know each other while serving on the boards and committees of professional associations and societies. These situations often allow the development of trust between even the most determined competitors and can lead to mutually beneficial partnerships. Competing companies can strengthen each other in the area of customer service. Business is such that cooperating on one aspect of business while competing on others often makes good sense.

Getting a team in the door for a visit is just the first step. Preliminary work is needed to ensure an effective visit. Figure 3.3 is a checklist for planning a visit. Step 1 is to decide on the specific purpose of the visit and to write it down. When personnel from both companies understand the purpose of the visit, they play their individual roles more effectively. Step 2 involves letting the host company know what is needed from them and specifically what is to be observed. In Step 3, team members receive a briefing about the purpose, logistics, and agenda for the visit. During the briefing, team members are given specific assignments, so that they know exactly what their responsibilities will be. Step 4 involves arranging an outbriefing with appropriate representatives of the host company. During the outbriefing, team members ask any remaining questions and representatives of the host committee make final comments. Step 5 is a planning meeting held as soon as possible after the visit so that information is still fresh and team members are not yet bogged down with their normal duties. In this meeting, appropriate personnel begin developing a plan for putting to use what was learned during the visit.

Checklist for Visiting Teams

1. Decide in advance what you are looking for. Write it down.
2. Give the company you will be visiting advance notice of specifically what you hope to accomplish.
3. Brief team members on the purpose of the visit and make specific assignments.
4. Arrange an outbriefing for asking questions and discussing concerns with representatives of the host company.
5. Organize a planning session based on what was learned as soon as possible upon returning to your company.

FIGURE 3.3 Good planning makes visiting teams more effective.

Figure 3.4 is an example of a purpose paper developed by the CEO of Southtron Company for members of a team planning to visit IN-TECH, Inc. The paper contains a simply stated but clear purpose as well as specific assignments for the team members. It is obvious that this CEO has done some preliminary work and knows about the various components of IN-TECH's customer-service plan. This purpose paper makes clear to team members why they are visiting IN-TECH and what each person is supposed to accomplish. This document would also be provided to the CEO of IN-TECH and her host personnel.

SELF-EVALUATION

Self-evaluation is an effective way to identify customer-service problems. Self-evaluation amounts to looking at your company from the perspective of a customer. There are several ways to do this. Some of the most effective include the following, as summarized in Figure 3.5.

1. *Call your own office.* If you want to know what customers hear when they call you, call your own office. Can you get through with no delay, or

Purpose Paper:
Visit to IN-TECH, Inc.

Purpose of the Visit

A team from *Southtron Company* will visit *IN-TECH, Inc.* on January 15th. The purpose of the visit is to observe *IN-TECH's* customer-service plan in action and to use what is learned to implement a similar plan at *Southtron Company.*

Specific Assignments

- Jones: Telephone component
- Matthews: Email/webpage
- Johnson: Receiving/recording verbal feedback
- Gonzalez: Self-evaluation component
- Van Thieu: Written feedback component
- Washington: Questionnaires
- Isoruka: Focus groups

FIGURE 3.4 A purpose paper can improve the effectiveness of a company visit.

Self-Evaluation Checklist

✓ Call your own office.
✓ Visit your company's website.
✓ Use your company's products.
✓ Lodge a complaint with your own company.

FIGURE 3.5 Self-evaluation strategies allow you to see your company through the eyes of the customer.

do you get caught up in the time-consuming machinations of a complicated computerized answering system? Does your secretary answer in a positive, friendly, helpful manner? Is it difficult to leave a message on your voice-mail? If any aspect of calling your office is even slightly problematic for you, it will be even more so for your customers.

2. *Visit your company's website.* Websites can be mixed blessings. On the one hand, they enable a company to be quickly and easily available to customers 24 hours a day, 7 days a week. On the other hand, if they are not well designed and frequently updated, they can drive customers away. Is your company's website easy to access? Does the website have a customer-friendly feel to it or is it off putting? Does the website portray the image your company wants to project, or another, less desirable image? Is it easy to find the information you need? What would make the website more customer friendly?

3. *Use your company's products.* If you want to know how customers will react to your company's products, use them yourself. If you work for Ford Motor Company, drive a Ford—but not just any Ford. Drive the model your plant manufactures. If your company produces products that are used by individuals, use them, and do so with a critical eye. What do you like about the product? What don't you like? What would make this product better?

4. *Use your company's services.* If your company provides services to customers, make use of them yourself, and do so with a critical eye. One of the authors once made major changes to the registration process after standing in line at his college to register for a course. In another case, the CEO of a construction company made several significant changes to his company's processes after having his home remodeled by one of his own crews. The same principle applies to any organization that provides services. If you want to experience what customers experience, use the services yourself.

5. *Lodge a complaint.* What actually happens when a customer registers a complaint? The best way to find out is to make one yourself and see what happens. Write a letter or send an email, and see how the company responds. Depending on the size of your company, you might need to use a different name. Remember when using this strategy that the purpose is to perceive your company from the vantage point of a customer, not to play "gotcha" with any individual employee.

CUSTOMER INTERVIEWS

You can learn a lot about customers' needs and preferences by just sitting down and talking with them. There are two approaches to conducting customer interviews: the "walk-in" approach and the "by-invitation" approach. The walk-in approach is used when a customer takes the initiative to make an appointment or just drops in. In either case, the customer has a problem he wants to discuss. The meeting is useful for more than just hearing the complaint. Once you have utilized the appropriate assertive listening strategies, you transform the meeting into an interview. This is done simply by asking the customer if he would mind answering some questions to help you improve customer satisfaction.

Safety Engineering & Manufacturing, Inc.
Customer-Interview Guide
Product: Chemical Protective Suit

This guide is provided to help you prepare for your interview with an SEM, Inc. representative on the following date: _____.

Please be prepared to provide your "likes and dislikes" for each of the following components of the SEM *Chemical Protective Suit:*

- **Components**
  - Suit
  - Coveralls
  - Hood
  - Gloves
  - Boots
- **Attributes of each component**
  - Comfort
  - Fit
  - Weight
  - Ease of dress/undress
  - Cost
  - Effectiveness of protection provided
  - Ability of employees to work while wearing

FIGURE 3.6 Preparation will minimize the time commitment of customers.

The by-invitation approach differs primarily in that the company takes the initiative to ask the customer for an interview. The customer is given advance notice of the types of questions the company representative plans to ask (see Figure 3.6). This gives her an opportunity to prepare, thereby making better use of everyone's time. Another important difference between the walk-in format and the by-invitation format is the number of customers that are interviewed. With the walk-in format, customers are taken one-by-one, as they appear. In essence, the walk-in customer is told, "You were coming in anyway. Since you are here, why not help us out by answering some questions?" With the by-invitation format, the number of interviews arranged depends on how many are necessary to adequately cover all segments of the company's market(s). For example, an engineering firm that provides services in more than one engineering field should interview several customers in each field. If a manufacturer produces more than one product, customers from all market segments should be interviewed. A comprehensive sample will yield the most accurate results.

Regardless of whether the interview is initiated by invitation or on a walk-in basis, its purpose is to collect information to be analyzed further and acted upon appropriately by company officials.

Who Should Conduct Interviews?

The effective conducting of customer interviews is both an art and a science. Many factors can affect the results of an interview, including:

- the objectivity of the interviewer
- the product or service knowledge of the interviewer
- the social skills of the interviewer
- the physical appearance of the interviewer
- the relative status of the interviewer (vis-a-vis the customer)
- the sensitivity of the interviewer to nonverbal cues
- the listening skills of the interviewer
- the questioning skills of the interviewer

Objectivity is an important characteristic of a good interviewer. If a company representative is biased or too invested in the company, product, or service to listen to constructive criticism, he will be an ineffective interviewer. For this reason, some companies choose to contract with professional third-party consultants to conduct customer interviews. This approach can relieve concerns about objectivity. If the product or service is at all complex, however, a third-party consultant may not have the technical knowledge to ask the right questions or to ask follow-up questions for clarification.

The issue of *product or service knowledge* can be important in deciding who should conduct customer interviews. If the product or the service involves a relatively simple consumer item, a third-party consultant who has been properly prepared should be able to conduct the interviews. However, as mentioned above, if the product or service is highly technical and complex, a third-party consultant is likely not to be an effective interviewer. In such cases, a company representative should be given the training necessary to conduct interviews.

The *social skills* of the interviewer can significantly affect her ability to establish a positive rapport and sense of trust with customers during interviews. Interviewers must be open minded, flexible, sensitive, and reassuring. There is disagreement as to whether these characteristics, often called "people skills," are the result of nature or nurture. Although it is true that some people seem to have a natural ability to deal effectively with people and others don't, it is equally true that even those who don't can improve through training and experience. An introvert who finds it difficult to deal with people may never be as strong an interviewer as an extrovert who enjoys the process. But an introvert can become good enough to conduct an effective interview—and that is as good as he needs to be in this case. Training related to conducting customer interviews is covered in Chapter 5.

The physical appearance of the interviewer, particularly in the area of dress, can have a positive or negative effect on the results of the interview. The key to appropriate appearance is striking the right balance between the professionalism expected of an interviewer and the normal dress of the customer. For example, an interviewer dressed in an expensive Saville Row suit with diamond rings on his fingers and Italian loafers on his feet might generate resentment in a middle- or lower-income customer. On the other hand, if the interview is with the CEOs of several Fortune 500 companies, an expensive suit and stylish footwear might be in order. What generally works best is dressing one notch above the customers' typical business apparel. You want to select clothing that projects an image of competence, confidence, and professionalism. You do not want to turn the interview into either a fashion show or a misguided attempt at dressing down. You are dressed just right for an interview if customers don't even notice your clothing and, as a result, can focus on your questions.

Relative status of the interviewer vis-a-vis the customer is critical when conducting interviews. The CEO of a company should not interview customers who are clerks in their jobs. In the same vein, clerks should not interview CEOs. People who conduct interviews should be at least equal to those they interview in both level of education and professional status; slightly higher is also acceptable. Too great a difference between the perceived status of the interviewer and that of the customer can produce inadequate results. We are more comfortable talking to people we can relate to, and we can relate to people with whom we share common ground. This rule of thumb also applies when hiring a consultant to conduct interviews.

Sensitivity to nonverbal cues is also an important skill for people who conduct interviews, a skill that can be learned. People communicate both verbally and nonverbally. Although most are aware of what they express verbally, many are not aware of what they express nonverbally. Tone of voice, steadiness of voice, posture, facial expressions, eye contact, proximity, and hand movements are all forms of nonverbal expression that give cues to the speaker's state of mind, emotional state, comfort level, and so on. It is a mistake to ascribe a specific meaning to any given nonverbal gesture, however. Rather, the key is to look for congruence or noncongruence between what is said verbally and what is communicated nonverbally. If the two do not seem to agree, there is a problem. For example, if a person insists that a certain thing happened or that a certain person said something, but is nervous and cannot bring himself to make eye contact, he is probably not telling the truth—or there are omissions in his account of events. When noncongruence exists between what is said verbally and what is expressed nonverbally, it is best to tactfully point out the disparity. When noticing noncongruence, one might simply ask, "Would you like to give that some more thought before giving me your final answer?"

To conduct effective customer interviews, it is necessary to be a good *assertive listener.* This topic was covered earlier. *Questioning skills* are also necessary, because it is important to have thorough, accurate customer information before taking any action. To get quality information, it is sometimes necessary to ask clarifying questions. When this is the case, the interviewer should give the customer a chance to completely state her case before asking questions and should never interrupt. The interviewer can make a mental note of any inconsistencies noted and wait until the customer has reached a stopping point. Unless there is a definite need for a "yes" or "no" answer, open-ended questions are best; they typically yield much more and much better information. Most questions can be framed in an open-ended format by restating them to begin with the phrase "Tell me about. . . . " For example, if a customer complains about rudeness on the telephone, a closed-ended question would be: "Did our employee hang up on you?" A question framed in this way is likely to elicit a one-word response. The customer is likely to give more information, some of which might be valuable in getting to the bottom of the problem, if he is asked an open-ended version of the question, such as: "Tell me how our employee handled your telephone call?"

Although third-party consultants can be hired to conduct *by-invitation* interviews, we recommend using company representatives whenever possible. Company officials who are not naturally good interviewers can and should be given sufficient training to correct this shortcoming. Even the best-prepared consultant is not as likely to have the feel for the company's product, service, or procedure that a company representative has. Moreover, some of the best customer interviews are initiated on a *walk-in* basis, in which case using a consultant is not an option.

Preparing for Customer Interviews

A customer interview is not a passive undertaking in which the information provided is simply collected as given. Rather, it is a systematic process whereby specific information in predetermined categories is sought. Of course it is easier to prepare for *by-invitation* interviews, but it is possible to prepare for the *walk-in* type as well, by undertaking the preparation explained in this section and having it available in the event a customer walks in.

The first step in preparing for an interview is to develop an interview checklist similar to the customer-interview guide presented in Figure 3.6.

Safety Engineering & Manufacturing, Inc.
Interview Checklist
Product: Chemical Protective Suit

Ask the customer to rate the quality of each attribute as excellent, acceptable but could be improved, or unacceptable. Ask for specifics about how the product could be improved. Repeat for each respective component.

Components

Suit, coveralls, hood, gloves, and boots

Attribute Ratings	Excellent	Needs Improvement	Unacceptable
• Comfort			
• Fit			
• Weight			
• Ease of dress/undress			
• Cost			
• Protection provided			
• Workability			

Comments:

____

____

____

____

FIGURE 3.7 Preparation will improve the quality of the interview.

Figure 3.7 is an interview checklist developed to collect specific information about the chemical protective suit from Figure 3.6. This is a product-oriented checklist. Figure 3.8 shows a service-oriented checklist.

The customer-interview guide in Figure 3.6 would be provided to the customers to be interviewed at least a week, and preferably two weeks, in advance of the interview. This is important because the customer might need time to collect feedback from users of the protective suit. Notice in Figure 3.6 that the

Safety Engineering & Manufacturing, Inc.
Interview Checklist
Services

Ask the customer to rate the quality of each service provided and to explain how the service could be improved.

Services	Excellent	Needs Improvement	Unacceptable
• Telephone			
• Website			
• 1-800-hotline			
• Walk-in (no appointment)			
• Product delivery			
• After purchase support			
• Personnel (interaction)			
• Billing			

Comments:

____

____

____

____

FIGURE 3.8 It is important to improve services as well as products.

various attributes listed apply to each component of the protective suit. The interview checklist in Figure 3.7 differs from the customer-interview guide in that it is prepared from the perspective of the person who will conduct the interview. As in Figure 3.6, the attributes listed must be applied to each component of the protective suit.

Figure 3.8 contains an interview checklist that covers the types of services provided by the designer and manufacturer of the protective suit (Safety Engineering & Manufacturing, Inc.). Customers are asked to rate the quality of each type of service and to comment on how it might be improved. Checklists and guides such as the ones in these examples can improve the quality of the information collected during interviews. They also minimize the length of the interview, because they help interview participants be well prepared.

Interview Report

As soon as an interview or a series of interviews is concluded, the interviewer should develop a report. It is important to develop the interview report while the information, observations, and impressions from the interview are still fresh. Any product attributes or services that need improvement should be noted, with the recommended improvements summarized. Copies of the report should be distributed to all stakeholders in the company for appropriate action. In addition, it is important to establish a tracking process so that needed improvements are not put aside and forgotten in the rush to conduct daily business.

FOCUS GROUPS

A focus group is a group of people convened, along with a facilitator, for the purpose of discussing specific topics. In the present context, focus groups consist of customers. The facilitator may be a company representative or a third-party consultant. The topics participants focus on, in this context, are customer needs and attribute preferences. Focus groups typically consist of 8 to 12 participants. They typically meet in a facility that is out of the way of traffic and other distractions, and that can accommodate group interaction, flip charts, marker boards, laptop computers, and other materials and equipment. Meetings should last no more than half a day. They typically run from two to three hours.

Focus groups can be used to provide feedback about product attributes and services or to provide input when a concept or product is still in the development stage. For example, a focus group could be used to provide feedback about the chemical protective suit in Figure 3.7 or about the company's services in Figure 3.8. A focus group could also be used to gather input about the desired attributes of the chemical protective suit while it is still in the development stage. Another focus group could be used for attribute testing during the development process. "Attribute testing" involves customers in trying out a product during the development of the preproduction prototype, to determine how they respond to its various attributes. The value of using focus groups for attribute testing is that the input allows the company to make revisions in the product's design before putting it into production.

Characteristics of an Effective Facilitator

- Is objective about the topic of discussion
- Has strong people skills
- Is tactful
- Knows when to stimulate discussion and when to shut it off and move on
- Can listen to a variety of opinions and categorize or summarize
- Can referee disagreements without being disagreeable
- Knows how to draw out reluctant participants
- Knows how to control participants who try to dominate the discussion
- Is able to paraphrase input and repeat it back succinctly
- Is able to ask open-ended questions for clarification and can sense when clarification is needed
- Can sense when to remain quiet and when to speak up
- Can build group consensus

FIGURE 3.9 The choice of facilitator is important to focus-group success.

Choosing the Facilitator

Who should facilitate focus-group meetings? The answer to this question is not as simple as it might seem. There are really just two options, a company representative or a third-party consultant. There are advantages and disadvantages with both options. Before getting into the relative merits of these two options, however, you need to know the characteristics needed by the facilitator, whether an employee or a consultant. These characteristics are summarized in Figure 3.9.

Typically, third-party consultants are people with specialized training and experience in facilitating focus groups. Such people should have the characteristics listed in Figure 3.9. Don't assume, however, that all consultants possess all these characteristics. Before hiring one, ask for a demonstration or some other proof of actual ability. The company may have individuals on staff who have the characteristics listed in Figure 3.9, but this often is not the case. Of course, company representatives can be trained to

Advantages/Disadvantages
Focus-Group Facilitator Options

Desired Characteristic	Company Representative	Third-Party Consultant
■ Depth of knowledge	X	
■ Direct observation	X	
■ Objectivity		X
■ Experience		X*
■ Facilitation skills		X*
■ Openness of participants		X
■ Responsiveness to questions	X	

*Typically favors the third-party facilitator

FIGURE 3.10 Company representatives and third-party consultants have inherent advantages and disadvantages.

facilitate meetings, and they should be if they are going to be called upon to do so. Some characteristics, such as strong people skills, tact, and objectivity, are difficult—some would say impossible—to teach, however. This brings up the nature versus nurture debate, which never really ends. One thing is clear, though: Certain personality types make better facilitators than others, regardless of training and experience (something that could be said about any occupation or profession).

Beyond the general characteristics that are desirable in a facilitator, there are several additional specific characteristics that can make the choice of one facilitator in a given situation better than another. Figure 3.10 shows the relative advantages and disadvantages of the two facilitator options (company representative or third-party consultant) when these additional specific characteristics are factored into the decision. The figure assumes that the individuals representing both options are equal in terms of the general characteristics desired of all facilitators.

Company representatives serving as facilitators generally have a greater *depth of knowledge* about products and services than third-party facilitators do. This can be an important advantage because it allows the facilitator to probe more deeply into what customers really want. It also allows the facilitator to ask questions for clarification and then to pose follow-up questions when customers respond. Using company representatives as facilitators also adds the advantage of *direct observation*. You can tell more about what a customer really thinks and feels by observing him directly than by listening to a third party describe the interaction. Direct observation allows the company representative to read body language, listen to the tone of voice, and watch for other cues that can be helpful in determining the importance of certain feedback to customers. Do participants feel strongly about a given issue or is it just a minor consideration? The best way to answer this question is through direct observation.

In spite of these important advantages, there are several disadvantages associated with using company representatives as facilitators—that is, there *are* advantages in using a third-party facilitator. The most important of these is *objectivity*. Because an outside facilitator is by definition a disinterested third party, she is more likely to be objective about customer feedback than a company representative would be. If a company facilitator is biased, he might lead the discussion away from a topic that to him is sensitive, thereby robbing the company of valuable customer information. A third-party facilitator, on the other hand, is likely to probe such topics more deeply.

Closely related to objectivity is the issue of *participant openness*. If the customers who participate in the focus group know the company facilitator (and sometimes even if they don't), they might be reluctant to speak their minds frankly out of fear of damaging the relationship. This is much less likely to happen with a third-party facilitator. Some people find it easier to open up to a complete stranger who is willing to listen than to family members or close friends. One of the authors once overheard an individual on an airplane bare his soul to his seatmate during a long flight. After the other passengers had exited the airplane, the author asked this individual, "Why would you tell such personal information to a complete stranger?" His response was simple. He said, "That's the whole point—I'll never see him again, so he's not a threat." This is the phenomenon that keeps psychologists and psychiatrists busy.

Two other characteristics—*experience* and *facilitation skills*—usually favor the third-party facilitator. This is only natural, because facilitating meetings is part of their profession. It bears repeating, however, that companies should never assume that a third-party consultant will automatically make an effective facilitator. Before hiring an outside facilitator, ask for a demonstration or some other verification of competence. In addition, build a requirement into the contract that the consultant provide a comprehensive summary report within a specified period of time.

The following rules of thumb can be applied when selecting a facilitator for customer focus groups:

- *Choose a company facilitator when*: (1) the product or services in question are clearly too complex for a third party to understand sufficiently;

(2) a company employee has solid facilitating skills; and (3) the customers who participate are not likely to have a relationship with the facilitator that might make them reticent to speak openly.

- *Choose a third-party facilitator when*: (1) the product or service in question is simple enough that the facilitator can ask probing follow-up questions; (2) facilitating skills are especially important and the company has no employee with the necessary skills; (3) the objectivity of company representatives is questionable; and (4) the participating customers have a close enough relationship with company officials that they might alter their feedback to protect the relationship.

WRITTEN FEEDBACK

The customer-feedback cards seen in restaurants and hotels have no direct application in an engineering, manufacturing, or construction setting, but the concept of soliciting written feedback from customers does. Although written customer feedback should be solicited by engineering, manufacturing, and construction companies, it works especially well for construction companies. Figure 3.11 is an example of a simple customer-feedback solicitation sent to a customer by email. The form identifies the relevant project and asks the customer two simple questions: How well were you served? and What can we do better?

The customer can answer conveniently by sending an email response. In this example, ABC Construction Company has finished the LMT Office Building and the owner has cut the ribbon. The owner is now in a position to take a retrospective look back on the entire project and give the construction company's CEO valuable feedback about the project. Written feedback should be solicited soon enough after the completion of a project that the customer's memory and impressions are still fresh, but not so soon that the customer has had insufficient time to even ponder the questions. Another key point about soliciting customer feedback is to keep it simple so that it takes as little time as possible on the part of the customer.

VERBAL FEEDBACK

Verbal feedback, also called "real-time feedback" (see Chapter 2), can be an effective way to collect useful customer information. In that chapter, we explained how to use verbal feedback to collect information for performance appraisals. In this section, a similar process is used to collect information about customer needs and attribute preferences. During the course of a project or contract, company officials receive a good deal of verbal feedback from

ABC Construction Company
Customer Feedback

On the *LMT Office Building* project, how well were you served?
What can we do better?

Your comments:

Name/Title

Date

Return to:
Marvin Bryant, CEO
ABC Construction Company
bryant@ABC.net

FIGURE 3.11 Customer feedback forms provide valuable information and are convenient for customers to use.

customers. Verbal feedback given in real time may be the most valuable feedback a company receives. Certainly it is the most current and timely.

It is unfortunate, then, that this valuable feedback is often lost in the rush to solve the problem and to keep the project on schedule. This need not be the case. Recording verbal feedback takes little time if companies develop a special form for doing so, such as the one in Figure 3.12. This form may be made available in hardcopy form, electronically, or both. Regardless of the format chosen, verbal feedback should be summarized when received, tied to a given project, and acted upon. By maintaining records of real-time customer

XYZ, Inc.
Customer Feedback Record

Project: ______________________________

Date: ______________________________

Feedback from: ______________________________

Recorded by: ______________________________

Record customer feedback in this space:

____
____
____
____
____
____
____
____

FIGURE 3.12 Verbal feedback should be recorded and used.

feedback tied to a given project, the company gains two valuable capabilities. First, it can analyze records to identify trends. If customer problems of the same type are found to occur frequently enough to establish a trend, they could identify a companywide or departmentwide problem. Second, it can tie negative or problematic employee behaviors to a given project. Do these behaviors occur more frequently on certain types of projects than on others? If this is the case, the company should investigate and determine why. At the very least, the information allows the company to predict potential problems, and to take steps to prevent them in the types of projects in question.

QUESTIONNAIRES

Questionnaires can be effective tools for collecting customer information, but only if they are properly designed and used. There are a number of factors that should be considered when designing a questionnaire (see Figure 3.13). A design process that fails to accommodate these factors is likely to produce a questionnaire that yields invalid customer information.

Design Factors for Questionnaires

- "Why" limitations
- Respondent bias
- Validity
- Meaningfulness
- Reliability

FIGURE 3.13 These factors must be considered when developing a questionnaire.

"Why" Limitations

Effective customer questionnaires are designed to help companies identify information of three types: (1) performance quality—How are we doing in meeting your needs for products or services?; (2) customer preferences—How important are certain product attributes or service characteristics to you?; and (3) comparison analysis—How are we doing compared to our competitors? What questionnaires cannot do, however, is explain *why* customers think or feel the way they do in these areas. This "why limitation" is important to note, because it represents a fundamental weakness of the questionnaire method of data collection. When the question to be asked of customers is "why," focus groups and interviews are more effective methods. To illustrate this point, consider the example of the customer who claims he prefers a competitor's product to your company's model. It is critical to understand why he prefers the competing model, but it is difficult to determine this using a questionnaire. In a focus group or interview, on the other hand, the facilitator could ask follow-up questions that would allow the customer to explain why.

All of this is to say that when it is important to understand why a customer responds in a given way, the questionnaire is the wrong tool to use. If it is important to know only "what" information such as customer preferences, however, a well-designed questionnaire can be an effective tool.

Respondent Bias

Respondent bias describes the condition that exists when an individual's responses to questionnaire items are influenced by factors unrelated to the product or service in question. Respondent bias can have either a favorable or an unfavorable leaning. Both leanings are equally undesirable. Bias in favor of a company's product or services is still bias, and biased information is inaccurate. It is not dependable when making critical decisions about improving customer satisfaction.

Respondent bias can be introduced by many factors. The following is a noncomprehensive list of factors that might introduce respondent bias in a questionnaire. The more effectively these factors are accounted for during the design phase of a questionnaire, the more effective the instrument will be in yielding accurate, dependable information.

- *Suggestive or leading wording in the instructions.* Wording in the questionnaire's instructions that is suggestive or leading can bias respondents' answers. For example, the authors once were asked to critique a questionnaire prepared by an engineering and manufacturing company for distribution to its customers. The instructions at the beginning of the questions read in part as follows: "We need your help. Our company has been nominated for a customer-service award to be given in January. In order to win this award, we must document customer satisfaction with our products. This survey is sent to you, one of our valued customers, for that purpose. . . ."

  You can easily see how such a statement might bias the answers of respondents. The instructions, rather than asking for objective, unbiased information, attempt to make respondents part of the team, so to speak, in pursuing the award. The authors recommended that this statement be dropped from the instructions, to delete the reference to the award.

- *Length of the questionnaire.* Customers faced with the prospect of completing a long, complex instrument might resent the intrusion on their time and, as a result, give unfavorably biased answers. Convincing customers to complete a questionnaire is difficult enough under the best of circumstances. Every effort should be made in the design of the questionnaire to keep it short.

- *Timing of the questionnaire.* In general terms, the best time to email or mail a questionnaire to customers is when the product or service in question is still fresh in their minds. Waiting too long can introduce "memory bias;" a condition that occurs when customers skew their answers either because they simply don't remember accurately or out of frustration at not remembering. "Intrusion bias" is another timing-related bias that occurs when the customer receives a questionnaire at a particularly intrusive time, such as over the weekend or on a holiday. It is recommended that questionnaires be sent to arrive on a Monday or Tuesday and never on holidays.

- *Uncontrollable biases.* Biases in the first three categories can be at least partially controlled. Those in this category cannot be, except through the sampling process. Some common sources of uncontrollable bias are as follows:

1. A respondent once had a bad experience with an employee in your company.
2. A respondent once had a bad experience with your company's products or services.

3. Someone the respondent knows had a bad experience with an employee, product, or service of your company and talked to the respondent.
4. A respondent read or heard something negative about your company or its products or services.
5. A respondent resents being asked to complete a questionnaire no matter how well designed it may be.

The key to minimizing the impact of uncontrollable biases (and minimization is all that can be done) is to make sure that the customer sample to which the questionnaire is sent represents a broad cross section of the overall customer base. This broad cross section is known as the "sample." The larger and more representative the sample, the less impact uncontrollable biases have on the results.

Validity

Several types of validity are associated with the design of questionnaires. In the current context, however, only two of them are relevant: construct validity and sample validity. Both should be considered and accommodated when designing a questionnaire.

Construct Validity

A questionnaire has construct validity when it actually measures what it is intended to measure—this concept applies both to the questionnaire as a whole and to each individual item. Construct validity is determined primarily by the wording of the question and the measurement method provided for responding to the question. Determining validity is an inexact science; the most effective way is to convene a focus group of customers and pilot test each questionnaire item. Participants complete each item one at a time and discuss with the facilitator their opinions concerning its validity. The more representative of the overall customer base this focus group is, the more valid the questionnaire is likely to be. The key in achieving construct validity is to eliminate ambiguity. Ambiguous questions run the risk of soliciting misleading answers, because each respondent may interpret the question differently. For example, consider the following questionnaire item:

Please rate the quality of our Model XYZ computer mouse:

_____ Excellent

_____ Good

_____ Fair

_____ Poor

This question is not likely to have construct validity because it is too ambiguous. Customers might differ as to how they define quality or what they see as excellent, good, fair, or poor. If a customer responds "Fair," what does that mean? What is wrong with the product that the customer views it as being only fair? To reduce ambiguity, the question might be rewritten as follows:

Please check the responses which best represent your opinions about our Model XYZ computer mouse:

_____ Comfortably fits the hand
_____ Is not comfortable in the hand
_____ Clicks easily
_____ Is difficult or inconvenient to click
_____ Is durable/low maintenance
_____ Needs too much maintenance

The second formulation of the question is less ambiguous than the first. The responses checked by customers will provide solid information on which to base improvements. Consider another example, from the field of construction:

Please rate our remodeling and renovation services:

_____ Excellent
_____ Good
_____ Fair
_____ Poor

Again, this formulation of the question is ambiguous. What services is the company asking about? What is meant by excellent, good, fair, and poor? More specificity is needed. Such a question should probably be broken up into several questions relating to specific remodeling and renovation services. For example, consider the following question:

Please indicate which of the following statements best describe your opinion of (the remodeling/renovation service in question). Choose all that apply.

_____ Our personnel listened to your ideas and based the final plans on your stated needs and preferences.
_____ Our personnel did not listen to your ideas, needs, and preferences as much as you would have liked.
_____ Our personnel provided you with an accurate estimate of the total cost of the renovations/remodeling and stuck to the estimate.
_____ Our personnel surprised you during the course of the project with costs that were not part of the original estimate.

There would need to be a number of additional questions, but put in this format there would be less ambiguity, and the construction company would have useful information to work with.

Sample Validity

A valid sample of a company's customer base is one that represents all segments of the overall base. For example, if a company's products are used throughout the United States, a sample taken only from the Northeast would not be a valid sample. When deciding who should receive questionnaires, a company has only two proper choices: (1) send the questionnaire to all customers (usually not a realistic option), or (2) select a representative sample of the overall customer base. Some companies invalidate the customer survey process by opting for convenience rather than sample validity. It might be more convenient to send questionnaires to just local customers or some other easily identifiable subset, but taking such an approach undermines the validity of the process.

Meaningfulness

Meaningfulness is another factor that can enhance or limit the usefulness of the questionnaire method of data collection, depending on how effectively it is accommodated in the design of the instrument. Responses to questionnaire items indicate the respondent's preferences, but what about the meaningfulness of those preferences? Sometimes what the company thinks is important may be less important to the customer. Consequently, when a respondent indicates either satisfaction or dissatisfaction with a given attribute, that is only half the story. The other half involves the relative importance to the customer of that attribute. A company would want to concern itself more with an attribute that produces dissatisfaction and is important to the customer than one that produces dissatisfaction but is not important to the customer. A company can waste a lot of time and resources making improvements to attributes that do not matter to the customer, when that time and those resources—both of which are always limited—could be directed at making improvements to higher-priority attributes.

For example, an automobile manufacturer designed a questionnaire to collect customer feedback about certain new amenities that had been added to one of its top-of-the-line models. One question asked respondents to rate the ease of operation of a new "back massage" vibrating system built into the driver's seat. The system is supposed to relax the driver and massage away the tensions of life while he is driving. The majority of customers indicated varying levels of dissatisfaction with the controls. Generally, the controls were found to be too complex and inconvenient. Redesigning the control mechanism for future models would be an expensive undertaking. Based on

Customer Questionnaire
Model XL-5 Lawnmower

Instructions

Step 1: Using the "Satisfaction Scale" provided on the left below, indicate your level of satisfaction with each of the attributes listed by placing the appropriate number in the blank to the left of the attribute. **Step 2:** Using the "Importance Scale" provided on the right below, indicate how important this attribute is to you by placing the appropriate number in the blank to the right of the attribute.

Satisfaction Scale					Importance Scale				
Not Satisfactory		Fairly Satisfactory		Very Satisfactory	Not Important		Fairly Important		Very Important
1	2	3	4	5	1	2	3	4	5

_____ Engine power _____
_____ Ease of starting _____
_____ Adjustability of wheel height _____
_____ Mulching feature _____
_____ Blade speed controls _____
_____ Adjustability of handle angle _____
_____ Access to fuel tank _____

FIGURE 3.14 Questionnaires should determine how important an attribute is to customers as well as their satisfaction with the attribute.

the customer feedback, should the manufacturer invest the time, energy, and money necessary to redesign the system's control mechanism? On the surface, customer feedback would appear to justify the expense. After all, the majority of respondents were unhappy with the control mechanism.

Fortunately for the company, an engineer on the design team suggested that a sample of respondents be resurveyed to determine how meaningful the issue of the control mechanism was to them. From this additional input, the company learned that although customers were almost uniformly unhappy with the control mechanism, few were interested in using the seat vibration system anyway. Customers had given their feedback about the controls for the vibration system because they had been asked for it, but in reality the system was a low-priority item for them. Based on this feedback, the company decided that rather than undertake an expensive redesign project, it would simply remove the vibration system from future models.

To avoid wasting limited resources improving attributes that are not important to customers, companies can include meaningfulness questions or codes in their questionnaires (see Figure 3.14). Notice in this figure that respondents indicate not just their level of satisfaction with each attribute, but also how important the attribute is to them.

Reliability

A reliable questionnaire repeatedly produces consistent results over time. The reliability of a questionnaire is a function of the effectiveness of its design and the validity of the samples. In other words, a reliable questionnaire is one that asks the right questions of the right people in the right way at the right time—not an easy accomplishment.

Summary

1. In the language of customer service, customer wants and preferences are often referred to as "needs." Almost all customer needs can be placed in one of the following categories: policy/system/process/procedure needs, people needs, and value needs.

2. Policies, systems, processes, and procedures control how companies interact with customers and how they deliver products and services to other companies. People needs grow out of the fact that customers want to deal with employees who are courteous, respectful, responsive, knowledgeable, and caring. Value needs are summarized in the "value equation," which has the following components: product quality, service quality, people quality, image quality, selling-price quality, and overall-cost quality.

3. Identification of customer needs should begin with an internal review of warranty records, customer-service records, input from frontline personnel, and professional involvement. These are all sources of potentially valuable customer information.

4. To maximize their ability to identify customer needs, employees should be taught to be assertive listeners. Assertive listening involves applying the following strategies: adopt an affirming attitude, block out distractions, maintain eye contact, do not react or deny, ignore the customer's attitude and concentrate on content, ask questions for clarification but do not interrupt, and paraphrase what was said.

5. A visiting team is a group of representatives from one company that visits other companies to observe best practices. The purpose of a visiting team is to observe what other companies do particularly well in the area of customer service and to discuss the observations with representatives of the company in question. The problem with this strategy is that companies are often reluctant to invite a competitor in to observe. This problem can be overcome by visiting companies in another, noncompeting region and by visiting companies in other lines of business. Also, the CEO can gain information through his counterparts in other companies with whom he has trusting relationships.

6. Self-evaluation for engineering, manufacturing, and construction companies is the equivalent of "ghost shopping" for retail businesses. Some retail businesses have their employees shop in their own stores to see first-hand how customers are treated. Although this approach does not translate exactly to engineering, manufacturing, and construction companies, it can be replicated sufficiently through self-evaluation. Self-evaluation methods include the following: calling your own office, visiting your company's website, using your company's products, using your company's services, and filing a complaint with your company.

7. The customer interview can be an excellent way to collect accurate customer information. When a customer comes to you with a complaint, listen and take the appropriate action. Then, turn the session into an interview by asking the customer open-ended questions about your company's products and services. Deciding who should conduct customer interviews is both an art and a science. The following characteristics of the potential interviewer all affect the decision: objectivity, social skills, product or service knowledge, physical appearance, relative status vis-a-vis the customer, sensitivity to nonverbal cues, listening skills, and questioning skills.

8. A focus group is a group of 8 to 12 people convened, along with a facilitator, for the purpose of giving input or feedback about selected topics. They typically meet in a facility that is out of the way of traffic and other distractions and that will accommodate group interaction, flip charts, marker boards, laptop computers, and other materials and equipment. Meetings typically last no more than half a day. A company representative makes the best focus-group facilitator when the following conditions exist: the product or services in question are too complex for a third-party facilitator to understand; a company representative has solid facilitating skills; and the participating customers are not likely to have a relationship with the facilitator. A third party facilitator is called for when the following conditions exist: the product or services in question are simple enough to be understood by a third party; the company has no one with the necessary skills; the objectivity of company representatives is questionable; and customers have a close enough relationship with company officials that they might alter their feedback to protect the relationship.

9. Written feedback can be a valuable form of customer information, but it must be solicited. A well-designed form that can be mailed, emailed, or given directly to customers can be a helpful tool for collecting written feedback.

10. Verbal feedback, also called real-time feedback, is very useful if it is recorded and acted on. The key is to have a standard form readily available to all employees so that they can record any verbal feedback they receive.

11. Surveys using questionnaires can be effective in collecting customer information, but they do have limitations. They are better for collecting "what" rather than "why" information. When designing a questionnaire, it is important to consider the following factors and account for them in the design: respondent bias, validity, meaningfulness, and reliability.

Key Phrases and Concepts

Assertive listening
Customer interviews
Customer needs
Customer-service records
Focus groups
Frontline personnel
Internal reviews
People needs
Physical appearance of the interviewer
Policy/system/process/procedure needs
Product or service knowledge
Professional involvement
Questionnaires
Relative status of the interviewer
Reliability
Respondent bias
Self-evaluation
Sensitivity to nonverbal cues
Social skills
Validity
Value needs
Verbal feedback
Visiting teams
Warranty records
"Why" limitations
Written feedback

Review Questions

1. List the three major categories of customer needs.
2. List five customer problems or needs associated with each of the three major categories of customer needs.
3. How might a company go about conducting an internal review to identify customer needs?
4. Explain the concept of assertive listening and how it can be used to identify customer needs.

5. What is a visiting team? How can a visiting team be used to identify customer needs?
6. How might a company use the concept of self-evaluation to identify customer needs?
7. Describe how a company should decide who will conduct customer interviews.
8. What is a focus group? How can companies use focus groups to identify customer needs?
9. Explain how a company might go about choosing a facilitator for its focus groups. When is it advisable to choose a company facilitator? When is it advisable to choose a third-party facilitator?
10. How can companies collect written feedback from customers?
11. How can companies collect verbal or real-time feedback from customers?
12. Explain the "why" limitations of questionnaires.
13. What is respondent bias? What can companies do to minimize respondent bias when using questionnaires to collect customer information?
14. List at least four examples of uncontrollable bias that might affect a customer's response on a questionnaire.
15. Explain the concepts of validity and reliability as they relate to the design and use of questionnaires.

ECS APPLIED: IDENTIFYING WHAT CUSTOMERS WANT AT DIVERSIFIED TECHNOLOGIES COMPANY

In the last installment of this case, David Stanley and his vice presidents at DTC took several specific actions to implement ECS. They revised job descriptions and performance-appraisal forms to include customer-service expectations, and they, along with all supervisors, were trained in the effective use of performance appraisals for recommending employee rewards and recognition. In this installment, Stanley and his vice presidents update each other on their efforts to formally identify customer needs and preferences.

David Stanley opened the weekly meeting of his executive management team by asking for updates from each of DTC's three vice presidents concerning their efforts to identify customer needs in their divisions. Stanley had asked the vice presidents to coordinate the following efforts within their respective divisions: a comprehensive internal review, customer interviews, focus groups, and a questionnaire. "Let's begin with engineering this morning," said Stanley. Meg Stanfield, vice president for engineering, was ready and eager. Stanley could tell that she had really bought into the ECS concept and was pursuing its implementation with all of her considerable talents.

"The internal review was very revealing," announced Stanfield. "I don't know why we haven't done this before now. But what I enjoyed most were the customer interviews and the focus groups. I facilitated both personally. David, that video training package you purchased on how to facilitate meetings is great. I didn't know there was so much involved in facilitating meetings, but there is, and—if I do say so myself—I'm getting pretty good at it." Stanfield went on to explain that the questionnaire for her division had been dropped in the mail that very morning. "We decided to use mail because we had more reliable postal addresses than we had email addresses." "Good report, Meg," said Stanley. He was always fascinated by how quickly Meg Stanfield could take the bull by the horns when given a project. "If only I had 10 more just like her," thought Stanley.

"Why don't you go next Tim?" asked the CEO. Tim Wang, vice president for manufacturing, gave a report that was almost an exact copy of Stanfield's, except that his questionnaire had been emailed the day before. Listening to Wang's report, Stanley silently marveled at the enthusiasm of this vice president. Both he and Stanfield were "can-do" types who always could be counted on to get the job done. Wang had used his director of manufacturing to conduct one-on-one interviews with customers, but he had conducted the focus groups himself. Both managers had completed the video seminar on facilitation before doing so.

Now it was Conley Parrish's turn to report. Stanley had noticed during the first two reports that his vice president for construction had seemed ill at ease. Over the last few weeks, he had also noticed some reluctance on Parrish's part to provide strong leadership in his division for implementing ECS. Conley was a good construction manager, but he did not have the enthusiastic "can-do" attitude of the other two vice presidents. "How are things going in the construction division, Conley?" asked Stanley. "I haven't made as much progress as Meg and Tim. That parking deck we are building over on Raleigh Boulevard has kept me and my folks tied down for the last month. I'll get to the ECS stuff as soon as I get the parking-deck project back on track." "Interesting," thought Stanley. "All of the weekly reports show that the parking deck is on schedule, maybe even a little ahead." Stanley decided to let it go for the moment, but he made a mental note to talk privately with Parrish at some point.

"How about dissenters?" asked Stanley. "Are we having problems?" He looked directly at Parrish when asking this question. The vice president for construction averted his eyes. Both Stanfield and Wang reported some foot dragging in their divisions, but nothing serious. "I have a couple of people who will need some one-on-one time," offered Wang. "But I think they will fall in line eventually." "Same here," said Stanfield. "How about in your division, Conley," queried the CEO. "I'm going to have some resistance. Most of my key people don't see the point." "Maybe Meg and Tim can offer some ideas as to how to bring them along," offered Stanley. "No need," said Parrish. "I can handle it." Stanley adjourned the meeting with an uneasy feeling about his vice president for construction. He made a mental note to take him to lunch later that week.

DISCUSSION CASES

The following cases provide examples of how the various concepts presented in this chapter might play out in actual companies. The cases are provided to prompt discussion, give the reader a feel for the types of problems confronted in the workplace, and reinforce the ECS concept in question.

CASE 3.1 I Wonder If This Is Happening to Our Customers?

The CEO of an engineering firm was out of town and needed to talk with someone in his home office. He was due in a meeting in just 30 minutes and had left an important file on his desk. "No problem," he thought. "I'll just call my secretary and have her fax me what I need." The CEO had recently directed that his company get rid of its old switchboard system and install a computerized "electronic receptionist." The new system, which replaced the human receptionist, was a marvel of telecommunications technology. According to its manufacturer, the new system would allow telephone calls to be answered and properly routed 24 hours a day, 7 days a week. Callers who know the extension of the person they need to talk to are instructed to punch it in on their touch-tone phone. Don't know the extension? No problem—just punch in the first four letters of the last name. What could be easier or more efficient? At least that is what the CEO had thought, but he made the mistake of not testing the system himself.

When the CEO called his office, his secretary was away from her desk helping another secretary learn to use a software upgrade, but she had set her telephone lines to forward automatically to another secretary in a nearby office. The CEO's call was forwarded to the designated station without a hitch. Unfortunately, that secretary had gone home unexpectedly to deal with a minor family crisis. The CEO got a recorded message. "No problem," thought the CEO. "I'll just call one of my engineers." After placing several calls and getting nothing but recordings, he remembered that all of the engineers were in their weekly staff meeting at the moment.

The CEO's next thought was to punch in the first four letters of the last name of any employee. Anybody could go to his office, pick the file up off of his desk, and fax the needed papers to him—provided that he could find someone who would actually answer the telephone. After several dead-end attempts, the now irritatingly serene voice of the computerized receptionist saying, "At the tone, please leave a message," was making the CEO frantic. "I need to talk to a human being!" Resolved to try one more time before giving up in complete exasperation, the CEO punched in the extension of the mailroom. When the company's mail clerk actually answered the telephone, the CEO was euphoric. As instructed, the mail clerk retrieved the file and faxed the needed papers to the CEO.

The much-relieved CEO participated in the meeting as if all was well with the world, but in reality he could focus on nothing but his computerized telephone system. "How many customers or potential customers have gotten hung up in the system and just given up?" he wondered to himself. He didn't know the answer to his question, but he did know one thing. As soon as he got back to his office, changes were going to be made. From now on callers would be able to talk to a human being.

Discussion Questions

1. Have you ever had trouble getting through a computerized telephone system when calling a company? Explain.
2. How did this encounter affect your opinion of the company?

CASE 3.2 The CEO's Secret Weapon

Meg Parsons has been the receptionist in the main lobby of S. A. Technologies, Inc. for more than 15 years. During this time, she has sought no internal promotions, preferring instead to remain in what her fellow employees view as a low-pay, low-prestige position. What they don't know is that Parsons is the CEO's secret weapon. Jane Patrillo became CEO of S. A. Technologies on the same day Meg Parsons was hired as the company's receptionist. The first person Patrillo asked to meet with was Parsons.

Earlier in her career, Patrillo had spent a lot of time in company lobbies waiting to meet with potential customers. There were always representatives of other companies also waiting, many of whom were her competitors. She noticed that they spent time between appointments chatting with the receptionist and with each other. Patrillo was amazed at the information these representatives would share during their chats. They talked about what they liked and did not like about the company they were calling on, as well as who they liked and did not like. After one particularly revealing episode of what Patrillo thought of as "lobby talk," she told herself, "If I'm ever a CEO, I am going to have my own personal set of ears in the lobby."

When Patrillo became CEO of S. A. Technologies, the first person she met with was the company's new receptionist, Meg Parsons. Patrillo explained that she wanted Parsons to listen to customers' lobby talk and pay special attention to any complaints or problems customers might be discussing about the company's products or services. Over time, Parsons became so good at recognizing customer issues that Patrillo began to pay her an end-of-year bonus. Because the bonuses were dependent on the relative value of the customer information Parsons provided, they varied in their amounts. There were years in which Meg Parsons's bonus was more than her annual salary. This is why she never pursued another higher-paying, more prestigious job within S. A. Technologies.

The steady stream of solid customer information that Parsons has provided to Patrillo over the years has allowed the CEO to make improvements both large and small. Although Patrillo implemented procedures for systematically collecting customer information from S. A. Technology's other frontline personnel, the CEO still thinks of Meg Parsons as her "secret weapon."

Discussion Questions

1. Have you ever sat in the lobby of a company waiting to see someone? Explain.
2. What types of things did you hear about the company from other people in the lobby?

CASE 3.3 I Can't Believe We Made This

Alex Washington is a former Marine, an ex–college football star, and an age-group record holder in a regional triathlon. Still an avid fitness enthusiast, Washington was delighted when he was selected from among more than 100 applicants to be the new CEO of Fitness Manufacturing, Inc. (FMI). A mechanical engineer by trade, Washington holds patents on several popular weight-training devices. FMI manufactures fitness equipment for both commercial and home use. Seldom have a company and a CEO been better suited to each other.

Washington had been the CEO of FMI for less than a week when he instituted a program that encouraged employees to use the company's products. He installed an on-site fitness center and started a "30/30" program, in which employees could use 30 minutes of company time and 30 minutes of their own time to work out on FMI equipment. Because many FMI employees are fitness enthusiasts, the 30/30 program was popular from the beginning, and it still is. But there is a catch. Employees pay for the 30 minutes of company time they receive by using the company's equipment with a critical eye. Employees are encouraged to "think like customers" and to make suggestions for improving any aspect of the equipment they use.

To give the new program a good start, Washington was the first person to use the new fitness center after the ribbon-cutting ceremony. Halfway through his workout, Washington noticed a problem with the company's new upright bench-press machine. The machine was designed to ensure perfect form on the part of every person using it, a concept Washington planned to use as the company's guiding principle and chief marketing slogan. The new CEO noticed that the machine not only did not ensure perfect form, it didn't even allow it. No matter how he adjusted the seat and the bar, Washington could not quite complete a repetition through the full range of motion. To make matters worse, the machine put excess pressure on the user's rotator cuff, something that could easily cause injuries.

Washington immediately called a meeting of the engineering team that designed the bench-press machine and explained the problem. He began the meeting by saying, "Ladies and gentlemen, I can't believe we made this machine. How does a product go through design, prototype, and testing and then get manufactured with so obvious a flaw?" It turned out that none of the members of the design team, all excellent engineers with impeccable credentials, ever worked out. Obviously the new machine would have to be pulled from production and design changes would have to be made. "Well," thought Washington. "It can't be helped. All I can do now is make sure this type of thing doesn't happen again." To the designers he said, "Folks, I cannot make you work out. That is a personal decision. But I can do this, and I am going to. I can change your job descriptions to ensure that you personally use every fitness device you design while it is still in the prototype stage. Nothing you design in the future will go into production until you have used it yourselves."

Washington was as good as his word. By the end of the day, the job descriptions of all design engineers had been revised. To date, the faulty bench-press machine was the last piece of fitness equipment developed at FMI to go into production without first being used in the prototype stage by its designers.

Discussion Questions

1. Have you ever purchased a product only to discover a problem with it that is so obvious that you wonder if anyone ever tested it before it was produced? Explain.
2. How did this situation affect your image of the company that produced the product?

CASE 3.4 I See a Pattern Here

Cedar Tree Construction Company's CEO, Joyce Goldwyn, is a real stickler for customer satisfaction. She is a hands-on CEO who visits job sites personally and frequently. If there is a problem, Goldwyn wants it solved immediately. She got to the top in a business traditionally dominated by men by maintaining close working relationships with customers and by going the extra mile to guarantee their satisfaction.

One of Goldwyn's techniques is to solicit written input from customers exactly one week after a project has been completed. She has found over the years that if she asks for feedback earlier than a week after the ribbon-cutting ceremony, the customer is still too focused on moving in to respond adequately. On the other had, if she waits longer than a week, the customer is likely to be too busy conducting business to respond adequately. Goldwyn's most valid and most valuable feedback comes when she sends a customer feedback form exactly one week after completing a project.

Several years ago, Goldwyn's company won the contract to develop an entire subdivision of single-family residences. More than 100 houses were to be constructed over a period of two years. Following her standard procedure, one week after a new owner moved in, Goldwyn paid a personal visit to say hello and drop off a customer feedback form. She even gave each customer a self-addressed, stamped envelope, to increase the likelihood of a prompt response. This attention to detail paid off in an important and unexpected way for Goldwyn. The first five forms returned to her all arrived in the mail on the same day. They gave the company high marks overall, but complained about one thing—the air-conditioning system. All five had the same complaint: The air-conditioning system seemed to be overloaded. It simply would not cool the new house sufficiently.

Goldwyn immediately contacted her air conditioning and heating subcontractor and explained the problem. The subcontractor, in turn, contacted the manufacturer. Within just days, the manufacturer shipped new units to replace the faulty ones. The manufacturer discovered, based on Goldwyn's feedback, that an entire production run of air-conditioning units had been manufactured with the wrong size cooling mechanism. The manufacturer gave its subcontractor the production-run number for the faulty units and asked him to return any that were currently on hand waiting to be installed in new houses. There were ten. These were quickly replaced with properly manufactured units.

Because Goldwyn regularly and systematically solicited customer feedback, she was able to identify and correct a customer-service problem that could have damaged her company's reputation as well as that of her air conditioning and heating subcontractor and the equipment manufacturer. Rather than unknowingly allowing a major problem to develop, she actually satisfied customers.

Discussion Questions

1. Have you ever purchased a faulty product only to get the runaround when you try to return it? How did this response make you feel as a customer?
2. Would you do business again with a company that refused to correct problems? Discuss the reasons for your answer.

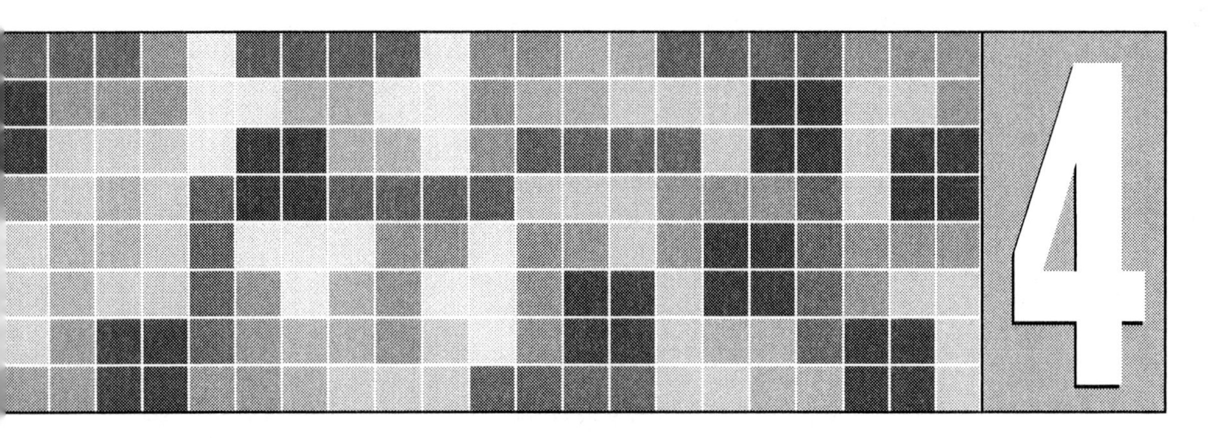

Benchmark the Company's Processes

If you want to thrive in a competitive marketplace, identify the companies that are best-in-class at customer service. Then set your company on a course to surpass them. You cannot win consistently in the long run unless you can outperform the best.

GOALS

- Understand the relationship among Steps 3, 4, and 5 in the customer service model.
- Define "benchmarking."
- Explain the rationale for benchmarking.
- Explain the steps in the benchmarking process.
- Identify the obstacles to effective benchmarking.
- List benchmarking resources.
- Act on benchmarking data.

What does it take for an athlete to make it to the Olympics and win a gold medal? What does it take for a professional baseball team to make it to the World Series and win? What does it take for a doubles team to make it to Wimbledon and win? Although those of us who are not Olympic, World Series, or Wimbledon champions may not know the answers to these questions, you can bet that athletes training to be champions in these fields know. They know because they apply the concept of *benchmarking*. During the Winter Olympics one year, a cellular telephone company ran an interesting commercial in which two competing speed skaters on different sides of the world used the "memo" function on their cellular telephones during training to compare best times for a lap. The first skater, having turned in a personal-best time, pages his competitor with the note "Beat this," and gives his time for the lap. Immediately the second competitor gets back on his ice and betters the first skater's time. He then returns the page, showing his latest time and writing, "No, beat this." This commercial illustrates the concept of best-in-class benchmarking. The skaters know that to win in the global arena—in this case the Olympics—they will have to outperform each other, because in the field of speed skating, one or the other of them is probably the best in the world. The commercial also makes the point that the concept of "best" is fluid and everchanging—a good point for companies, as well as athletes, to remember. Continual improvement is essential in a competitive environment.

OVERVIEW OF BENCHMARKING

Benchmarking emerged as a continual improvement tool in the 1980s. As a concept, it takes the mystery out of competitiveness. Businesses that effectively benchmark their performance know quantitatively the difference between consistently winning and only sporadically winning, or worse yet, losing. Benchmarking is the best way for a company to know how well it must perform to be competitive in its markets.

Motorola is an example of a company that gained best-in-class status in part because of its decision to use benchmarking as a strategic tool. Even after achieving best-in-class status, Motorola continues to use the benchmarking process before implementing any new programs, products, or services. First the company asks, "Who is best-in-class with regard to the process in question?" Benchmarks are then set accordingly. In this way, new products, programs, and services introduced by Motorola are introduced at world-class levels.

Benchmarking is a necessary component in the implementation and ongoing application of ECS, but it is still misunderstood by many. As widely practiced as the concept is in today's international business arena, some people still view benchmarking as industrial espionage, cheating, or unethical. It is none of these. In fact, when practiced as explained in this chapter, benchmarking is always a *cooperative* undertaking, either within a given company or between organizations that knowingly and willingly share performance information.

THE RELATIONSHIP AMONG STEPS 3, 4, AND 5

It is important that readers understand the relationship among Steps 3, 4, and 5 of the ECS implementation model. In Step 3 (covered in Chapter 3), companies identify the product attributes and service characteristics that are most important to customers. In the step covered in this chapter (Step 4), companies benchmark the processes that produce or affect those attributes and characteristics. Then in Step 5 (Chapter 5), companies compare the performance of the processes in question against the benchmarks established in Step 4. These three steps, taken together, enable a company to know specifically how to satisfy customers better than any other competitor can.

BENCHMARKING DEFINED

To successfully undertake benchmarking, one company must be willing to learn from another.

> ***Benchmarking** is the process of comparing and measuring an organization's operations or its internal processes against those of a best-in-class performer from inside or outside its industry.*

Xerox is an example of a well-known company that uses benchmarking as a strategic tool to stay on top in its industry. In the late 1970s, Xerox was losing significant market share to Japanese competitors, who were producing excellent products and selling them at prices below Xerox's manufacturing costs. When Xerox executives began to study this untenable situation, they found that their company had nine times as many suppliers as a typical Japanese competitor and seven times as many manufacturing defects. Lead times for new products at Xerox were twice as long, and production setup times were five times as long as those of the Japanese competitors.

To fight back, Xerox implemented benchmarking in 1980. Its processes were benchmarked against the best in the world in and out of its industry. As a result of these efforts, Xerox was able to save itself. Today Xerox is a world-class company capable of holding its own against any competitor—foreign or domestic. Benchmarking has affected every facet of the company and remains a primary reason for the company's success.

Benchmarking involves finding the secrets for success for any given function or process so that a company can learn from the information—and improve on it. It helps a company narrow the gap between itself and the best-in-class performers without having to reinvent the wheel.

A distinction exists between benchmarking and competitive analysis. *Competitive analysis* involves comparing a competitor's product against yours. It compares the features, attributes, and pricing of the product. Consumers, as well as companies, perform competitive analyses—for instance, when they are trying to determine which brand of VCR, television, or automobile to purchase. *Benchmarking* goes beyond that—to comparing how a product is engineered, manufactured, constructed, distributed, and supported. Benchmarking is interested not so much in what the product is and what it costs, but rather in the underlying processes used to produce, distribute, and support it.

Perhaps the most important benefit of benchmarking is that it helps establish where improvement resources should be allocated. If, for example, it is discovered that three of five processes are nearly as good as those of the best-in-class performers, but two are significantly off the best-in-class mark, then the most benefit for the dollars invested will come from changing the weaker two to conform more nearly to the best-in-class. Less will be gained by drastically changing the processes that are already close to the best.

Key points to remember about benchmarking are as follows:

- Benchmarking is an increasingly popular improvement tool.
- Benchmarking concerns processes and practices.
- Benchmarking compares one company's process or practice with a target company's best-in-class process or practice.
- Benchmarking is a respected means of identifying processes that require major change.
- Benchmarking is carried out between consenting companies that may or may not be competitors.
- The goal of benchmarking is to find "secrets of success" and then adapt and improve on them for the company's own application.
- Benchmarking is equally beneficial for large and small companies.

RATIONALE FOR BENCHMARKING

The future for companies today seems far different from what it did after World War II. The first real questions regarding the ability of the United States to sustain its industrial leadership resulted from the oil crisis of 1974. By then, the United States had lost much of its lead in the commercial electronics business to Sony, Hitachi, and Panasonic, but the premiere industry in the United States, automobile manufacturing, seemed secure. When the oil embargo struck, however, Americans quickly traded their big, domestic cars for small, fuel-efficient Japanese models. When the embargo ended,

Americans continued buying Japanese cars because at that time they were of higher quality than their American counterparts. The Japanese quickly claimed about 30 percent of the U.S. automobile market (and possibly could have gained much more if it were not for voluntary restraints adopted out of fear that severe trade restrictions would be imposed by Washington). Following these events, American companies started waking up to the fact that the world was changing. Whole industries were moving from one part of the world to another, and most of that movement was to Japan. There was good reason to look to Japan and see what they were doing differently that reaped them such success.

What was learned, of course, was that by following the teachings of Deming, Juran, Ishikawa, Taguchi, Ohno, and other quality pioneers, Japan had developed superior practices and processes. These resulted in high-quality manufactured goods at competitive prices—everything from motorcycles, to cars, cameras, electronics of all kinds, and even shipbuilding. It took several years for American businesses to realize fully what had happened. For a long time, Western leaders attributed Japan's success to low labor costs, the Japanese work ethic compared with Detroit's, lifetime employment, and other factors. Such rationalizations clouded the real issue: the superiority of Japanese practices and processes. Now that industrial leaders worldwide are aware that better practices and processes can enhance competitiveness, it makes good business sense for organizations to determine where they stand relative to world-class standards, and what they must do to perform at that level. This is what benchmarking is designed to do.

Twenty years ago, benchmarking involved comparing American industries with Japanese industries. Today, benchmarking involves comparing a company with the best in the world. The best in the world may be in Japan, or it may be next door. It may be a direct competitor, or it may be in a completely different industry. In addition to companies all over the world emulating the best, customers all over the world are demanding the highest quality in the products they buy. Business as usual no longer works. Organizations must improve continually and forever, or they will be out of business soon and forever.

The rationale for benchmarking is that it counters the nonsensical situation of a company isolating itself and inventing a new process to improve a product or reduce its cost, when that process already exists. If one company has a process that is more efficient, it is logical for other companies to adopt that process. An organization might make incremental improvements to its process through continuous improvement, but it could take years to make a 4X improvement, and by then the competition would probably be at 6X or better. Benchmarking reveals which processes are candidates for continuous (incremental) improvement and which require major (one-big-step) changes. Benchmarking offers the fastest route to significant performance improvement. It can focus an entire organization on the issues that really count. Key points to remember about benchmarking as it relates to continuous improvement are as follows:

- Today's competitive marketplace does not allow time for gradual improvement in areas in which a company lags behind.
- Benchmarking tells a company where its practices and processes stand relative to the best-in-class and which processes must be changed.
- Benchmarking provides a best-in-class model to be adopted or even improved on.
- Modern customers are well informed and demand the highest quality and lowest prices. Companies have a choice: either perform with the best or go out of business.
- Benchmarking supports ECS by providing the best means for rapid, significant process and practice improvement.

BENCHMARKING PROCESS: HOW TO DO IT

The benchmarking process is relatively straightforward, but the steps must flow in a sequence. A number of variations are possible, but the general sequence is as follows:

1. Obtain management commitment.
2. Baseline your own processes.
3. Identify your strong and weak processes and document them.
4. Select processes to be benchmarked.
5. Form benchmarking teams.
6. Research the best-in-class companies.
7. Select candidate best-in-class benchmarking partners.
8. Form agreements with the benchmarking partner.
9. Collect data.
10. Analyze data and establish the gap.
11. Plan action to close the gap or surpass the benchmark.
12. Implement change.
13. Monitor the charge.
14. Update benchmarks and continue the cycle.

These 14 implementation steps are explained in order in the following sections.

Step 1: Obtain Management Commitment

Benchmarking is not something to be approached lightly. It requires a great deal of time from key people, and money must be made available for travel to the benchmarking partners' facilities. Time and money require management approval. You expect to gain information from your benchmarking partner, and they will expect payment in kind—namely, information from you about your processes. This exchange of information also must be authorized by management. Finally, benchmarking will likely result in replacing or making major changes in processes. Such changes cannot be made without management approval. Without a mandate from top management, there is no point in attempting to undertake a benchmarking process. If you cannot secure management commitment, you need proceed no further.

Step 2: Baseline Your Own Processes

If your company is involved in ECS, chances are good you have already done some baselining of processes, because this is necessary for any effective continuous improvement. The processes must be characterized in terms of capability, flow diagrams, and other factors. If this has not been done before, it must be done now. It is critical that you thoroughly understand your existing processes before you compare them with someone else's. Most employees think they know their company's processes well, but this is rarely the case without a deliberate process characterization. Also, if it has not been done already it is important that an organization's processes be completely documented for the benefit of everyone associated in any way with them.

Step 3: Identify and Document Strong and Weak Processes

Strong processes will not be benchmarked initially; continuous improvement techniques will be sufficient for them. Weak processes, however, become candidates for radical change through benchmarking, because incremental improvement would not be sufficient to bring them up to the level necessary in the timeframe required.

It can be difficult to categorize an organization's processes as weak or strong. A process that causes quality problems is an obvious choice for benchmarking. On the other hand, a process may be doing what is expected of it and as a result be classified as strong; it could be, however, that expectations for that process are not high enough. Another organization may have a much more efficient process, but you just don't know about it. No process should be considered above benchmarking, no matter how highly it is rated. Concentrate on the weak ones, but keep an open mind about the others. If research identifies a better process, add it to the list.

Above all, document all processes fully—even the strong ones. Keep in mind that just as you will be looking at your benchmarking partner's superior processes, they will be looking at your strong processes as well. Undocumented processes are not helpful to benchmarking partners. It is impossible to compare two processes unless both are fully documented.

Step 4: Select Processes to Be Benchmarked

When you have a good understanding of your own processes and the expectations associated with them, decide which ones to benchmark. Never benchmark a process you do not wish to change; there is no point in doing that. Benchmarking is not something you engage in simply to satisfy curiosity. The processes on a benchmark list should be those known to be inferior and targeted for change. Leave the others for incremental change through continuous improvement—at least for the time being.

Step 5: Form Benchmarking Teams

The teams that will do the actual benchmarking should include people who have input into the process, those who operate the process, and those who take output from it—the employees who are in the best position to recognize the differences between your process and that of your benchmarking partner. The team must include someone with research capabilities, to help identify a benchmarking partner. Every team should have management representation, not only to keep management informed, but also to build the support from management that is necessary for radical change.

Step 6: Research the Best-in-Class Companies

It is important that a benchmarking partner be selected on the basis of being best-in-class for the process being benchmarked. In practical terms, it is necessary that the potential partner be willing to participate. Because benchmarking examines processes, the best-in-class may be in a completely different industry. For example, an organization that manufactures copy machines might tend to consider potential benchmarking partners who are leaders in the copying industry. But if it is a warehousing process that is to be benchmarked, the company might get better results by looking at catalog companies that have world-class warehousing operations. If the process to be benchmarked is accounts receivable, perhaps a credit card company would be a good partner.

Because processes are shared across many industries, limiting research to companies in like industries might mistakenly disqualify the best opportunities for benchmarking. "Best-in-class" does not mean "best-in-industry." It just means "best" for the process in question—regardless of industry. If team members stay up to date with trade journals, they should be able to compile a good list of potential benchmarking partners. Research sources should include trade literature, suppliers, customers, Baldrige Award winners, and professional associations. The Internet offers a comprehensive supply of benchmarking information. As they carry out their research, team members will find that documentation about best-in-class processes becomes readily available.

Step 7: Select Candidate Best-in-Class Benchmarking Partners

When the best-in-class companies have been identified, the team must decide which among them it would prefer to work with. Consideration must be given to location and to whether the best-in-class is a competitor (because the team will have to share information with the partner). The best benchmarking partnerships provide some benefit for both parties. If the team can identify benefits for its potential partner, the linkage between the two companies will be easier to achieve. Even without that, most companies with best-in-class processes are willing—often eager—to share their insights and experience with others. Indeed, Baldrige Award winners are expected to share information with other U.S. companies.

Step 8: Form Agreements with the Benchmarking Partner

After the team has selected the preferred candidate, it contacts the potential partner to form an agreement covering benchmarking activities. It can be useful to have an executive from your company contact an executive from the target company, especially if there is an existing relationship or some other common ground. Identifying the appropriate contact person in the potential partnering company is often the most difficult part of the process. Professional associations can sometimes provide leads about people who would be in positions with the necessary authority.

After such a contact has been made, the first order of business is to determine the company's willingness to participate. If the potential partner is unwilling, the team must choose another candidate. With willing partners, an agreement can usually be forged without difficulty. The terms include visitation arrangements, limits of disclosure, and points of contact. In most cases, these terms are informal. Even so, care must be exercised not to require excessive obligations of either benchmarking partner. Make the partnership as unburdensome as possible.

Step 9: Collect Data

In this step, the team observes, collects information about, and documents those of the partner's processes that it had determined to study. In addition, it tries to determine the factors underlying the success of processes. Does the partner employ continuous improvement, for example, or employee involvement, use of statistics, or other strategies? Optimally, the process operators should talk directly with the partner's operators. It is important to come away with a good understanding of their process (in the form of a flow diagram) and its support requirements, timing, and control. The team should also try to gain some understanding of the preceding and succeeding processes, because if one is changed the others may require change as well. If the team knows enough when it leaves the partner's company to implement the process back home, it has learned most of what is needed. Anything less than this, and the team has more work to do.

While visiting a partnering company, try to get a feel for how they operate. Be receptive to new ideas that are not directly associated with the process in question. Observing another organizational culture can stimulate a wealth of ideas worth pursuing.

Step 10: Analyze the Data and Establish the Gap

With the data in hand, the team undertakes a thorough analysis and compares it with the data taken from its own process. In most cases, the team should be able to establish the gap (the performance difference between the two processes) numerically—for example, 200 pieces per hour versus 110, 2 percent scrap versus 20 percent, or errors in parts per million rather than in parts per thousand.

If the team concludes that without a doubt the partner's process is superior, other questions arise. Should we replace our process with theirs? What will it cost? Can we afford it? What impact will it have on adjacent processes? Can we support it? Only by answering these questions can the team determine whether implementation is possible.

Step 11: Plan Action to Close the Gap or Surpass the Benchmark

If the team concludes that a change to the new process is desirable, affordable, and supportable, it recommends adoption of the process. In most cases, implementation requires some planning to minimize disruption while the change is being made and while the operators are getting accustomed to the new process. It is important to approach implementation deliberately and with great care. This is not the time for haste. Consider all conceivable contingencies, and plan ahead so as to avoid them, or at least to be prepared for them. Physical implementation may be accompanied by training for the operators, suppliers, or customers. Only after thorough preparation should an organization implement the change to the new process.

The objective of benchmarking is to put in place a process that is best-in-class. If the team merely transplants the partner company's process, it may not achieve the objective, although improvements may result. To achieve best-in-class, an organization must surpass the performance of the benchmark process. It may not be possible to do this at the outset, but the team's initial planning should provide for the development work necessary to achieve it in a specified period of time.

Step 12: Implement the Change

The easiest step of all may be the actual implementation, assuming that the team's planning has been thorough and that execution adheres to the plan. New equipment may or may not be involved, and there may be new people—or more or fewer people—but there certainly will be new procedures that need to become routine. It should not be a surprise, therefore, if initial performance does not equal the benchmark. After people get used to the changes and the initial problems get worked out, performance should become close to that of the benchmarking partner.

Step 13: Monitor the Change

After the process is installed and running, performance should come up to the benchmark quickly. Then, continuous improvement should enable the organization to surpass the benchmark. None of this is likely to happen without constant monitoring. Never install a new process, get it on line and performing to expectations, and then forget about it. All processes need constant attention in the form of monitoring.

Step 14: Update Benchmarks and Continue the Cycle

As was explained in Step 11, the intent of benchmarking is not only to catch up with the best-in-class but to surpass them, thereby becoming best-in-class yourself. This is a formidable undertaking, because those with best-in-class processes are probably not resting on their laurels. They too are striving for continually better performance. However, you are now applying new eyes and brains to their process, and fresh ideas may well yield a unique improvement, vaulting your organization ahead of the benchmarking partner. Should that happen, your organization will be sought out as a best-in-class benchmarking partner by others who are trying to elevate their performance. Whether that happens, and even whether the benchmark is actually surpassed, the important thing is to maintain the goal of achieving best-in-class. Benchmarks must be updated periodically. Continue the process. Never be content with a given level of performance.

An important consideration, as you either achieve best-in-class or get close, is that resources have to be diverted to those processes that remain lowest in performance relative to their benchmarks. Let continuous improvement practices be applied to the best processes, and concentrate the benchmarking on the ones that remain weak.

OBSTACLES TO EFFECTIVE BENCHMARKING

Like most human endeavors, benchmarking can fail. Failure in any activity usually stems from inadequate preparation—not learning enough about the requirements, the rules, and the pitfalls. So it can be with benchmarking. This section presents some of the common obstacles to successful benchmarking.

Internal Focus

For benchmarking to work, you have to know that someone out there has a better process. If a company is internally focused (as many are), it may not even be aware that its processes are 80 percent less efficient than the best-in-class. An internal focus limits vision. Is someone better? Who is it? Some organizations don't even ask the questions. This is complacency—and it can destroy an organization.

Overly Broad Benchmarking Objective

An overly broad benchmarking objective, such as "Improve bottom-line performance," can guarantee failure. This may be the general reason for benchmarking, but the team needs a more specific goal—oriented not to the *what* but to the *how*. A team could struggle with the bottom line forever without knowing with certainty whether it had achieved success or failure. Benchmarking teams need a narrower target, such as "Refine the invoicing process to reduce errors by 50 percent."

Unrealistic Timetables

Benchmarking is an involved process, and it cannot be compressed into a few weeks. Consider four to eight months the shortest timeframe for an experienced team, with six to eight months the norm. A shorter schedule will force the team to cut corners, which can lead to failure. If you want to take advantage of benchmarking, be patient. Any project that goes on for more than a year should be checked, however. The team is probably floundering.

Poor Team Composition

Benchmarking teams must be composed of those employees who own the process, the people who use it day in and day out. These may be production-line operators or clerks. Management may be reluctant to take up valuable team slots with lower-level personnel when the positions could otherwise be occupied by engineers or supervisors. Certainly, engineers and supervisors should be involved, but not to the exclusion of the process owners. The process owners know best how the process really operates; they can most easily detect the often subtle differences between your process and that of the benchmarking partner. Teams usually are made up of six to eight people. Be sure the first members assigned are the operators; there will still be room for engineers and supervisors.

Settling for "OK-in-Class"

Too often organizations choose benchmarking partners who are not best-in-class, for one of three reasons:

- The best-in-class is not interested in participating.
- Research identified the wrong partner.
- The benchmarking company got lazy and picked a handy partner.

Organizations get involved in benchmarking when they decide that one or more of their processes is much inferior to the best-in-class. The intention is to examine that best-in-class process and adapt it to local needs, quickly bringing the organization up to world-class standards in that process area. It makes no sense to link with a partner whose process is just good. Although it may be better than yours, it would still leave your organization far below best-in-class. For the same amount of effort, an organization could make it to the top. Only if the best-in-class will not participate can accepting a second-best process be justified. Second-best should be used only if it is significantly superior to the company's own process.

Improper Emphasis

Benchmarking teams frequently get bogged down collecting endless data, and they often put too much emphasis on the numbers. Both data collection and the actual numbers are important, of course, but the most important element is the process itself. Gather enough data to understand your partner's process on paper, and analyze the numbers sufficiently to ascertain that your results can be improved significantly by implementing the new process. It is best to keep the emphasis on the process, with data and numbers supporting that emphasis.

Insensitivity to Partners

Nothing breaks up a benchmarking partnership more quickly than insensitivity. Remember that a partner is doing your organization a favor by allowing access to its process. The partner is donating the valuable time of its key people, and it is tolerating the disruption of its daily business routine. If you fail to observe protocol and common courtesy in all transactions, your organization runs the risk of being cut off.

Limited Top Management Support

This issue keeps coming up because it is so critical to the success of all stages of the benchmarking activity. Unwavering support from the top is required to get benchmarking started, carry it through the preparation phase, and finally secure the promised gains.

BENCHMARKING RESOURCES

A number of resources are available to help organizations with benchmarking efforts. They cover the spectrum from nonprofit associations to cooperative affiliations to for-profit organizations that sell information. In addition, of course, there are consulting firms with expertise and databases related to all aspects of benchmarking.

One of the most promising ventures is the American Productivity and Quality Center (APQC) Benchmarking Clearinghouse (123 N. Post Oak Lane, Houston, TX 77024; phone [713] 685-4657; fax [713] 681-5321), www.apqc.org. The APQC Benchmarking Clearinghouse assists companies and nonprofit organizations in collecting and disseminating best practices through databases, case studies, publications, seminars, conferences, videos, and other media.

A wide range of benchmarking information is available on the Internet. Just enter "benchmarking" or "process benchmarking" in your search engine, and you will be rewarded with more information than you can use. Internet resources range from articles on benchmarking to promotions for books and consultants. Colleges list their libraries' benchmarking-related content. A word of caution: Anyone can put anything on the Web without verification, so it is always wise to approach material from unfamiliar sources with a degree of skepticism. Despite this one caveat, we consider the Web to be a valuable benchmarking resource.

Trade and professional organizations are excellent sources of information on benchmarking. They can direct organizations to best-in-class practices, provide contacts, and offer valuable advice. Baldrige Award winners are committed to share information with other U.S. companies, and they hold periodic seminars for this purpose.

Trade literature publishes a wealth of relevant information, including lists of companies with best-in-class processes and practices. *Industry Week* is one example of an excellent source of benchmarking information. Companies such as Dun and Bradstreet and Lexis-Nexis maintain databases of potential benchmarking partners, which they share for a fee.

Consultants and universities that are engaged in benchmarking can help organizations get started by providing initial training, offering advice and guidance, and directing organizations to benchmarking partner candidates.

No matter what resources you utilize, be sure that any information obtained is current. The very nature of benchmarking makes yesterday's data obsolete. To achieve maximum benefit, organizations must be sure they are operating based on current information.

ACTING ON THE BENCHMARK DATA

At the conclusion of the benchmarking project, data analysis will have produced both quantitative and qualitative information. The quantitative information is effectively the stake driven into the ground as the point from which future progress is measured. It is also the basis for improvement objectives. Qualitative information relates to such things as personnel policies, training, management styles and hierarchy, quality maturity, and so on. This information provides insights into how the benchmarking partner got to be best-in-class.

Quantitative data are clearly the information sought and are always acted upon. Qualitative information may be even more valuable, however. It describes the atmosphere and environment in which best-in-class processes and practices can be developed and sustained. Do not ignore it. Study it, discuss it in staff meetings, and explore the possibilities of introducing changes based on it into your culture.

In terms of the process that has been benchmarked, if the partner's process is significantly superior to your own—and it should be, or it would not have been selected in the first place—you have to do something about implementing it. Perhaps you can modify your own process with some ideas picked up from benchmarking. Or more likely, you can adopt your partner's process, implementing it to replace yours. Whatever is indicated by the particular situation, take decisive action and get it done.

Summary

1. Benchmarking is a process for comparing an organization's operations or processes with those of a best-in-class performer. The objective of benchmarking is major performance improvement, achieved quickly. Benchmarking focuses on processes and practices, not products.

2. Benchmarking is carried out between consenting organizations. Benchmarking partners are frequently from different industries. Benchmarking must be approached in an organized, planned manner, with the approval and participation of top management.

3. Benchmarking teams must include those who operate the processes. It is necessary for the benchmarker to understand its own process before comparing it with another.

4. Because best-in-class status is dynamic, benchmarking should be seen as a never-ending process. Management has a key role in the benchmarking process, including committing to change, making funds available, authorizing human resources, being actively involved, and determining the appropriate level of disclosure.

5. The goal of benchmarking is to become the best-in-class, not simply improved. The intent of benchmarking may be either to replace an inferior process with one rated best-in-class, or to radically improve a process, bringing it up to best-in-class performance. In both cases, the additional goal is to surpass best-in-class.

6. A number of obstacles to successful benchmarking exist, including internal focus, overly broad or undefined objectives, unrealistic timetables, inappropriate team composition, failure to aim at best-in-class, diverted team emphasis, insensitivity toward the partner, and wavering support by top management.

Key Phrases and Concepts

Benchmarking
Benchmarking partner
Best-in-class
Continuous improvement
Customer focus
Internal focus
Key business process
Performance gap
Process baselining
Process capability
Process characterization
Process flow diagram

Review Questions

1. Define "benchmarking."
2. How do continuous objectives and benchmarking objectives differ?
3. List five factors that lead organizations to benchmarking.
4. Which processes should an organization concentrate on for benchmarking?
5. Why is it necessary that top management be committed as a prerequisite to benchmarking?
6. What are the reasons for characterizing and documenting an organization's processes before benchmarking?

7. Identify the critical members of the benchmarking team.
8. Why it is not enough to simply clone the benchmarking partner's process?
9. Explain the importance of linking the benchmarking objectives with the organization's strategic objectives.
10. How can the "not-invented-here" syndrome be a hindrance to benchmarking effectiveness?
11. List and discuss the eight obstacles to successful benchmarking.

ECS APPLIED: BENCHMARKING AT DIVERSIFIED TECHNOLOGIES COMPANY

In the last installment of this case, two of DTC's vice presidents—Meg Stanfield (engineering) and Tim Wang (manufacturing)—reported on the progress they had made in their respective divisions in identifying customer preferences with regard to product attributes and service characteristics. Both divisions had conducted internal reviews, customer interviews, and focus groups. In addition, both divisions had designed and distributed a questionnaire to solicit customer feedback. DTC's other vice president, Conley Parrish (construction), had failed to get his division engaged in a project to identify customer needs. David Stanley, CEO of DTC, is beginning to worry about Parrish's commitment to the implementation of ECS. In this installment, Stanley and his vice presidents discuss the company's benchmarking project.

"Where is Conley?" asked Meg Stanfield. "I can't remember a time when he missed a meeting." "I don't know," answered Tim Wang. "Now that you mention it, I haven't seen him much lately." "Nor I," added David Stanley. "In fact, he cancelled a lunch meeting with me earlier this week, and not until the last minute." The CEO asked the two vice presidents to wait another five minutes, but when Conley Parrish still had not arrived, he started the meeting. "Well, Conley is probably tied up at a jobsite. Let's go ahead and start."

Stanfield and Wang had prepared a PowerPoint presentation for the CEO, and they proceeded to deliver it jointly. The presentation explained the joint benchmarking project that was being undertaken by the engineering and manufacturing divisions. A key slide in the presentation dealt with key process areas in the two divisions and the primary processes within those areas. The slide contained the information shown in Figure 4.1.

Stanfield and Wang explained that they had found two different partnering companies for benchmarking the various processes. Finance and human resources processes were being benchmarked against the same processes at PrintCom, Inc., a partner in another industry. Engineering, production, and quality management processes were being benchmarked

Diversified Technologies Company
Engineering and Manufacturing Divisions

Key Process Areas with Corresponding Primary Processes

- Engineering
  - Research and development
  - Design
  - Product improvement
- Finance
  - Accounting
  - Accounts payable
  - Accounts receivable
- Human Resources
  - Recruiting/hiring
  - Compensation/benefits
  - Employee development
- Production
  - Procurement
  - Materials
  - Warehousing
  - Material control
  - Material preparation
  - Production control
  - Assembly
  - Integration and testing
- Quality Management
  - Incoming inspection
  - In-process inspection
  - Supplier auditing
  - Internal auditing

FIGURE 4.1 Sample slide showing key process areas.

against the same processes in Z-Tech Manufacturing, Inc. Both companies were being helpful and cooperative. The teams from DTC were working well with their counterparts from PrintCom and Z-Tech.

"Good report and good job," said David Stanley, obviously pleased with the progress the engineering and manufacturing divisions were making. "Any preliminary indications of where we are going to need the most work?" "In manufacturing, we will need to focus on procurement, production control, and assembly," responded Tim Wang. Meg Stanfield spoke for engineering. "In my division, we will have to focus on research and development. It is taking us twice as long as it takes Z-Tech to move new products out of the lab and into production." "Well, at least now we know what it takes to play and win in the big leagues," said Stanley.

After the meeting, David Stanley asked his secretary to block out an hour on his calendar for a meeting with Conley Parrish. "I would like to meet with him first thing in the morning. If that doesn't work, make it as soon as possible. Make sure you tell Mr. Parrish that it's important."

DISCUSSION CASES

The following cases provide examples of how the various concepts presented in this chapter might play out in actual companies. The cases are provided to prompt discussion, give the reader a feel for the types of problems confronted in the workplace, and reinforce the ECS concept in question.

CASE 4.1 Why Don't We Look Outside of Our Industry?

Walcom, Inc. is a diversified company with engineering, manufacturing, and construction divisions. As a result, its warehouse operations are complex. The company's warehouse personnel must manage, inventory, handle, and distribute parts and materials ranging from high-grade stainless steel to lumber and cement. The ability to move parts and materials into and out of the warehouse with maximum efficiency is critical to Walcom's competitiveness. When the CEO of Walcom, Inc. decided to benchmark its warehouse operations, his warehouse manager could not find a comparable company to approach. There were numerous manufacturing firms with excellent warehouse operations. There were also a number of construction companies that fit the bill. But he could find no company that, like Walcom, handled materials and parts from both industries as well as engineering support materials.

As he discussed the situation with Walcom's CEO, the warehouse manager seemed to have arrived at a dead end. It wasn't, he explained, that he didn't want to undertake the benchmarking project. Quite the contrary. He knew his operations were holding back Walcom's efforts to become more competitive. He just couldn't find a suitable benchmarking partner. After listening to his obviously frustrated warehouse manager for several minutes, the CEO asked, "Have you ever ordered anything from L.L. Bean?" The question was so unexpected and off the wall that the warehouse manager had a confused moment before he could answer. "Well, uh no, I guess I haven't." The CEO handed him a copy of Bean's latest catalog and said, "Flip through this for a few minutes while I go refill my coffee cup."

When he returned, the CEO asked, "Notice anything interesting about that catalog?" "Well," chuckled the warehouse manager, "they carry a great line of hiking shoes." "Yes they do," said the CEO. "They also carry backpacks, tents, canoes, skis, clothing, and numerous other items that are even more diverse than the parts and materials we carry in our warehouse. Let me ask you a question. If there are no benchmarking opportunities within our industry, why don't we look outside? I understand that L.L. Bean not only has a best-in-class warehouse operation, but that company officials are open to discussing their methods with others." Within an hour after finishing his meeting with Walcom's CEO, the warehouse manager had purchased an airline ticket for Freeport, Maine.

Discussion Questions

1. Have you ever dealt with more than one company to secure the same product (perhaps at different times)?
2. If so, compare the quality of the products and services of these companies. Which company is better and why?

CASE 4.2 The Benchmarking Project that Bombed

Electric Equipment Company (EEC) undertook its current benchmarking project with the intention of becoming best-in-class in its markets. That was almost 12 months ago. Now the CEO was conducting a meeting with his benchmarking team to determine what went wrong. Although EEC made improvements in the areas of product quality, technical support, and human interaction, the changes to date had been insufficient to move EEC into a stronger position vis-à-vis its competitors. Members of the benchmarking team were apprehensive. EEC's CEO was known as a tough and demanding executive who sets the bar high and expects results.

The CEO began the meeting by unveiling a chart listing the various steps for implementing a benchmarking project. "Here is what we did or, at least, what we say we did," said the CEO. "Take a few minutes to review this chart, and then let's discuss what went wrong." After a prolonged silence, broken only by the nervous clearing of throats and shifting in chairs, the team's captain finally broke the ice. "I think maybe our benchmarking objectives were too broad." The CEO nodded and asked, "Can you give me an example of what you mean?" "Sure. Look at the invoicing process as an example. Our objective for this process was to improve its efficiency by 50 percent. This was a good objective, but what did we really mean by it? If we had set more specific objectives for the invoicing process, we would be better able to monitor the level of improvement." "How would you break the invoicing objective into more specific objectives?" asked the CEO. After some discussion, participants settled on three specific areas of concern with regard to invoicing: accuracy, timeliness, and responsiveness to inquiries.

"How would you put these areas of concern into the form of measurable objectives?" asked the CEO. "Well, let's take invoice accuracy," offered one participant. "We could have set a specific objective in this area of 99 percent invoice accuracy. In fact, 99 percent accuracy is the performance benchmark we got from our partner company." "Why didn't we use the benchmark they gave us?" questioned the CEO. He saw the team's captain visibly wince at the question. Another participant offered, "Well . . . we weren't able to get a specific benchmark for timeliness and responsiveness. Without those other two benchmarks, all we could do was lump the three criteria together and apply what we thought was a reasonable percentage." "Now," said the CEO, "tell me about why we couldn't get the more detailed information."

The room fell silent, and participants took a sudden interest in looking at their hands or studying their notes—anything to avoid eye contact. After an uncomfortable silence, the team's captain decided that when caught in a corner, the best way to get out is to tell the truth. "We blew it. We forgot that our partners don't work for us. Looking back on it now, I'd have to say we were too demanding on the one hand, and insensitive to their time constraints on the other." "So I heard," said the CEO in a tone of voice that caused participants again to take an interest in studying their notes. Sensing their discomfort, the CEO said, "OK, we blew it. That's my fault, not yours. I got all of you into this without any preparation. I apologize. I did not call this meeting to assess blame. If there is anyone to be blamed here, it's me. What I want out of this meeting is an open, frank discussion about what went wrong with our benchmarking project. Then I want us to try again and get it right this time."

There were audible sighs of relief around the room, and within minutes participants were immersed in a good discussion of why and how the project had gone awry. After consolidating several points and refining some others, the team had a well-thought-out list of errors to avoid in the future. "Now we know what went wrong," said the CEO. "Look at the errors on this list, and I think you will agree that their root cause was insufficient preparation of team members. Consequently, I'm going to bring in a benchmarking consultant to get us ready to try again. I'll give him this list of problems before our first meeting so he will know exactly what went wrong on the first try. Then, while you team members are working with the consultant, I'm going to begin looking for another benchmarking partner. I don't think our first partner wants us back." This last comment was said with a smile, but the CEO could tell from the group's reaction that he had made his point.

Discussion Questions

1. Have you ever worked in a "partnership" situation in a joint project with personnel from another organization? Did all personnel on the team get along well? Explain.
2. How did interactions among team members from different companies affect the work of the team? Explain.

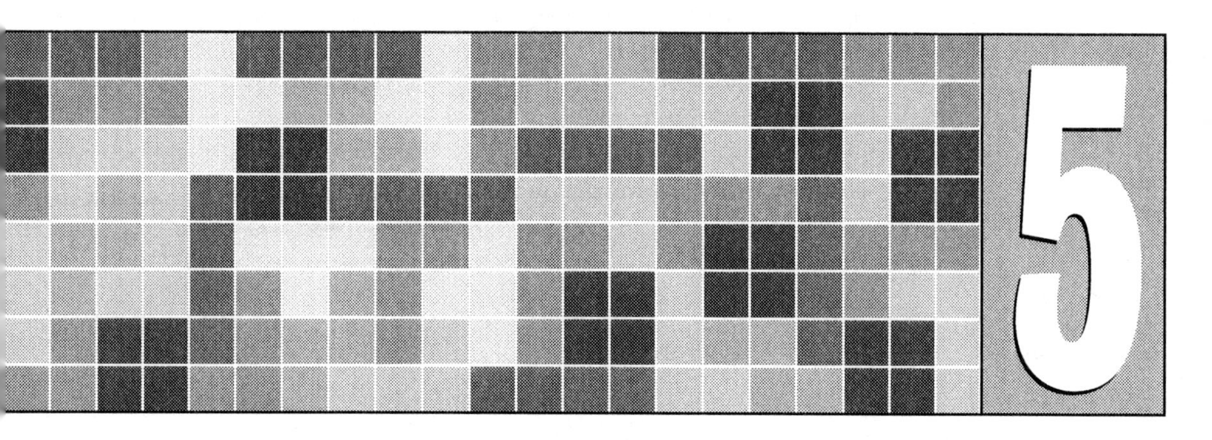

Compare Actual Performance Against Benchmarks, Identify Root Causes of Performance Problems, and Make Improvements

Once a company determines the level of performance necessary to be best-in-class, it should compare this benchmark to its actual performance. The difference between these two poles is the difference between world-class competitiveness and mediocrity.

GOALS

- Understand the relationship among Steps 3, 4, and 5 of the ECS implementation model.
- Explain the purpose and demonstrate the use of: pareto charts, cause-and-effect diagrams, check sheets, histograms, scatter diagrams, run charts and control charts, and other helpful tools.
- Explain the rationale for continual improvement.
- List and describe essential improvement activities.
- Name the components of a structure for improvement.
- Describe the importance of the scientific approach to improvement.
- Develop improvement plans.
- List common improvement strategies.

One of the basic tenets of ECS is management by facts. This is not in harmony with the capability so revered in North America and the West in general: the ability to make snap decisions and come up with quick solutions to problems in the absence of input beyond intuition, gut feel, and experience. Management by facts requires that each decision and each solution to a problem be based on relevant data and appropriate analysis. Other than in the very small business (where the data are resident in the few heads involved), most decisions and problems are complex, and the problems' root causes or the best-course decisions remain obscure until valid data are studied and analyzed. Collecting and analyzing data can be difficult. The performance-enhancement tools presented in this chapter make the task less difficult. Their use will help assure better decision making, better solutions to problems, and even improvement of productivity, products, and services.

THE RELATIONSHIP AMONG STEPS 3, 4, AND 5

It is important that readers understand the relationship among Steps 3, 4, and 5 of the ECS implementation model. In Step 3 (covered in Chapter 3), companies identify the product attributes and service characteristics that are most important to customers. In Step 4 (covered in Chapter 4), companies benchmark the processes that produce or affect those attributes and characteristics. Then in the step covered in this chapter (Step 5), companies compare the performance of the processes in question against the benchmarks established in Step 4. These three steps, taken together, enable a company to know specifically how to satisfy customers better than any other competitor can.

OVERVIEW OF PERFORMANCE-ENHANCEMENT TOOLS

Carpenters use an array of tools designed for very specific functions. Their hammers, for example, are used for driving nails and their saws for cutting wood. These tools and others enable carpenters to do their jobs. They are *physical* tools. Performance-enhancement tools enable a company's employees—whether engineers, technologists, production workers, managers, or office personnel—to do their jobs. Virtually no one can function in an organization that has embraced ECS without some or all of these performance-enhancement tools. Unlike those in the carpenter's toolbox, these are *intellectual* tools. They are not wood and steel to be used with muscle; they are tools for collecting and displaying information in ways that help the human brain grasp thoughts and ideas. When thoughts and ideas are applied to physical processes, better results are achieved. When they are applied to problem solving, continuous improvement, or decision making, better solutions and decisions are developed.

The tools explained in this chapter represent those generally accepted as the basic performance-enhancement tools. Each tool is some form of chart for the collection and display of specific kinds of data. These charts present the data in ways that make it meaningful and useful. The information can be used to solve problems, enhance decision making, keep track of ongoing work, even predict future performance and problems. Again, the value of the charts is that they organize data so that people can immediately comprehend the information. This would be all but impossible without these tools, given the amount of data available in today's workplace.

THE PARETO CHART

The *Pareto* (pah-ray-toe) *chart* is a useful tool for separating the important from the trivial. The chart, first promoted by Dr. Joseph Juran, is named after Italian economist/sociologist Vilfredo Pareto (1848–1923). He had the insight to recognize that in the real world just a few causes lead to the majority of problems. This is known as the "Pareto principle." Pick a category, and the Pareto principle will usually apply. For example, in a typical business, only about 20 percent of all possible problems produce 80 percent of customer complaints. Charts have shown that approximately 20 percent of the pros on the tennis tour reap 80 percent of the prize money and that

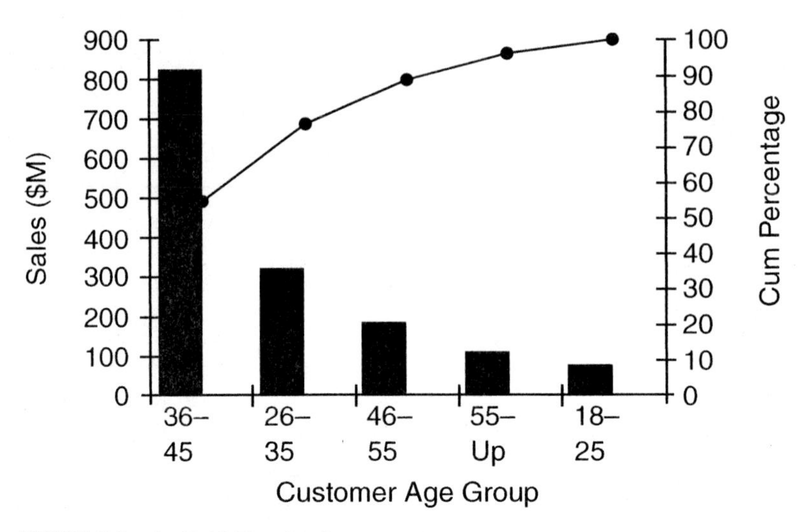

FIGURE 5.1 Swift V-12 sales by age group.

80 percent of contributions to churches in the United States comes from 20 percent of the church membership.

All enterprises—from giant corporations to the government—have limited resources. Because they are limited, the resources (time, energy, and money) need to be applied where they will do the most good. The purpose of the Pareto chart is to identify where the resources could best be applied to achieve the greatest performance improvements.

The Pareto chart in Figure 5.1 shows bars representing the sales of a particular model of automobile by age group of the buyers. The curve represents the cumulative percentage of sales and is keyed to the right-hand y-axis scale. The manufacturer has limited resources in its advertising budget, and the chart reveals which age group should most logically be targeted. Concentrating on the 26–45 age bracket will result in the best return on investment because 76 percent of the Swift V-12 buyers come from the combined 36–45 and 26–35 age groups. The significant few referred to in the Pareto principle are in the 26–45 group. The insignificant many are those under 26 and over 45.

Cascading Pareto Charts

You can cascade Pareto charts by determining the most significant category in the first chart, making a second chart related only to that category, and then repeating this as far as possible—to three, four, or even five or more charts. If the cascading is done properly, root causes of performance problems can be determined.

Consider the following example. A company that produces complex electronic assemblies was concerned about the cost of rework resulting from test failures. They were spending more than $190,000 per year on rework,

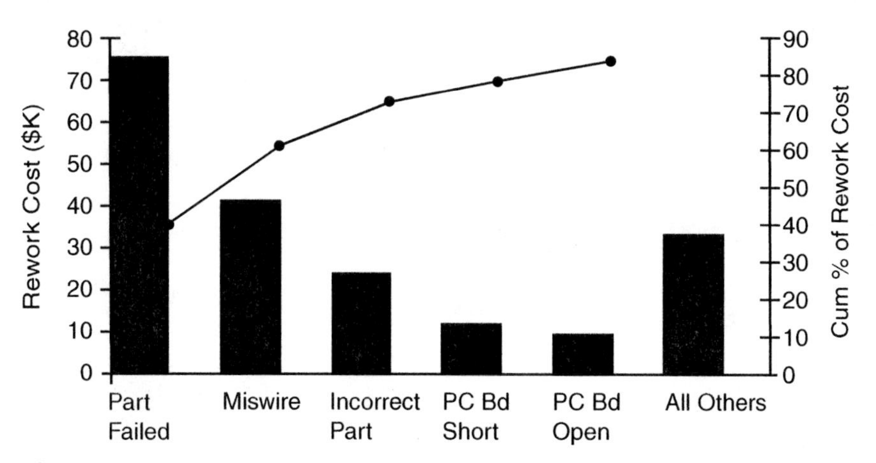

FIGURE 5.2 Top five defects by rework cost.

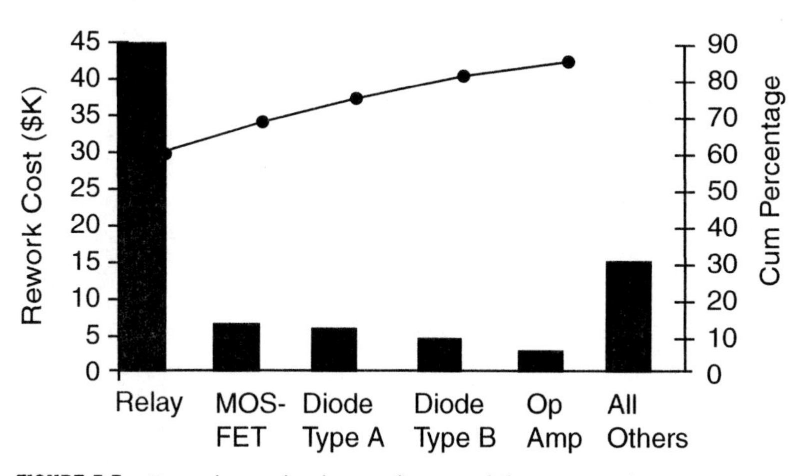

FIGURE 5.3 Rework cost by the top five part failure categories.

and that came directly out of profit. Benchmarking showed that this figure must not exceed $15,000. The test department formed a special project team to find the cause of the problem and reduce the cost of rework. The Pareto chart in Figure 5.2 showed them that about 80 percent of the cost was related to just five defects. All the others, and there were about thirty more, were insignificant, at least at that time.

The longest bar alone accounted for nearly 40 percent of the cost. If the problem it represents could be solved, the result would be an immediate reduction of almost $75,000 in rework costs. The team sorted the data again to develop a level-two Pareto chart, which focused on part types that might be a major contributor to the failures (see Figure 5.3).

Figure 5.3 clearly showed that one type of relay accounted for about 60 percent of the failures. No other part failures came close. In that case and at that time, the relay was the significant one, and all other parts were the insignificant many.

CAUSE-AND-EFFECT DIAGRAMS

A team typically uses a *cause-and-effect diagram* to identify and isolate the causes of a problem. This is an important first step in improving performance to benchmarked levels. The technique was developed by the late Dr. Kaoru Ishikawa, a noted Japanese quality expert. Consequently, the diagram is sometimes called an "Ishikawa diagram." It is also often called a "fishbone," because that is what it looks like.

The cause-and-effect diagram is the only performance-enhancement tool described in this chapter that is not based on statistics. This chart is simply a means of visualizing how the various factors associated with a process affect its performance. The same data could be tabulated in a list, but the human mind would have a much more difficult time associating the factors with each other and with the total outcome of the process under investigation. The cause-and-effect diagram provides an easily deciphered graphic view of the entire process.

Suppose an electronics plant is experiencing soldering rejects on printed circuit (PC) boards. To improve performance to the benchmark level, soldering rejects must be reduced by at least 75 percent. Engineers at the plant decide to analyze the process; they begin by calling together a group of people to get their thoughts. The group is made up of engineers, solder-machine operators, inspectors, buyers, production-control specialists, and others. All the groups in the plant that have anything to do with PC boards are represented, which is necessary to get the broadest possible view of the factors that might affect process output.

The group is told that the issue to be discussed is the solder defect rate and that the objective is to list all the factors in the process that could possibly affect the defect rate. The group brainstorms to generate a list of possible causes. The list is shown in Figure 5.4. It is a comprehensive list of factors in the PC board manufacturing process—factors that could *cause* the *effect* of solder defects. Unfortunately, the list does nothing in terms of suggesting which of the 35 factors might be major causes, which might be minor causes, and how they relate to each other. This is where the cause-and-effect diagram comes in. Ishikawa's genius was to develop a means by which these random ideas could be organized to show relationships that would help in making intelligent choices.

Figure 5.5 is the basic form of a cause-and-effect diagram. The spine points to the *effect*. The effect is the "problem" being addressed—in this case, machine soldering defects. Each of the ribs represents a cause leading to the

machine	solderability	operator
solder	conveyer speed	temperature
preheat	materials	parts
operator attitude	operator attention	flux
conveyer angle	wave height	cleanliness
age of parts	age of boards	part preparation
parts vendors	board vendors	type of flux
specific gravity	machine maintenance	training
skill	vibration	storage
instruments	lighting	calibration
handling	wait time	contamination
air quality	humidity	

FIGURE 5.4 Brainstormed list of possible causes for solder defects.

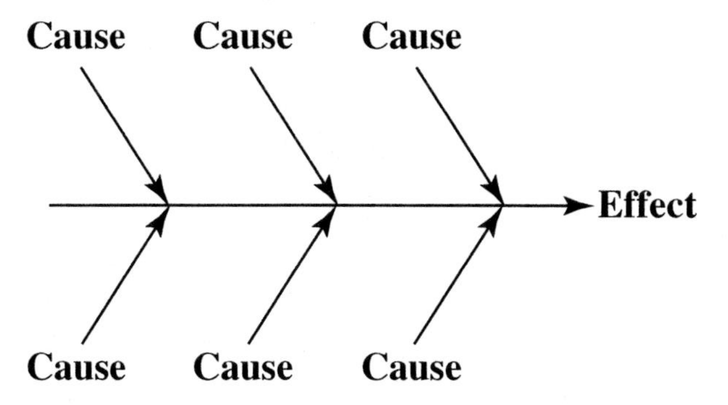

FIGURE 5.5 Basic fishbone diagram.

effect. The ribs are normally assigned to the causes considered to be *major factors*. Lower-level factors affecting the major factors branch off the ribs.

The group makes the following decisions in trying to determine the major factors causing the soldering rejects:

1. The solder machine itself is a major factor in the process.
2. The operators who prepare the boards and run the solder machine are also major factors.
3. Many items in the list, such as parts, solder flux boards, and so forth, can be collected under the word "materials," which also appears on the list. Materials constitute a major factor.

4. Temperature within the machine, conveyor speed and angle, solder wave height, and so on are really the *methods* (usually published procedures and instructions) used in the process. Methods constitute a major factor.
5. Many of the items in #4 are subject to the plant's methods (how-to-do-it) and measurement (accuracy of control), so measurement is a major factor, even though it does not appear on the list.
6. Cleanliness, lighting, temperature and humidity, and the quality of the air can significantly affect performance and thus the quality of output. This major factor can be called "environment."

The six major factors the group thinks might affect the performance of the machine-soldering process are, then: machine, operator, materials, methods, measurement, and environment.

After assigning the major causes, the next step is to assign all the other causes to the ribs they affect. For example, machine maintenance should be assigned to the Machine rib, because machine performance is obviously affected by how well the machine is maintained. Training should be attached to the Operator rib, because operators' training certainly affects their expertise in running the machine. In some cases, a cause noted on the list may appropriately branch not from the rib (major cause) but from one of the branches (contributing cause). For example, solderability (the relative ease or difficulty with which materials can be soldered) would branch from the Materials rib. An important cause of poor solderability is age of parts. So age of parts would branch not from Materials, but from Solderability.

Figure 5.6 shows the completed fishbone diagram. It presents a picture of the major factors that can cause solder defects and in turn the smaller factors that affect the major factors. Examination of the Materials rib shows that there are four factors directly affecting materials with regard to solder

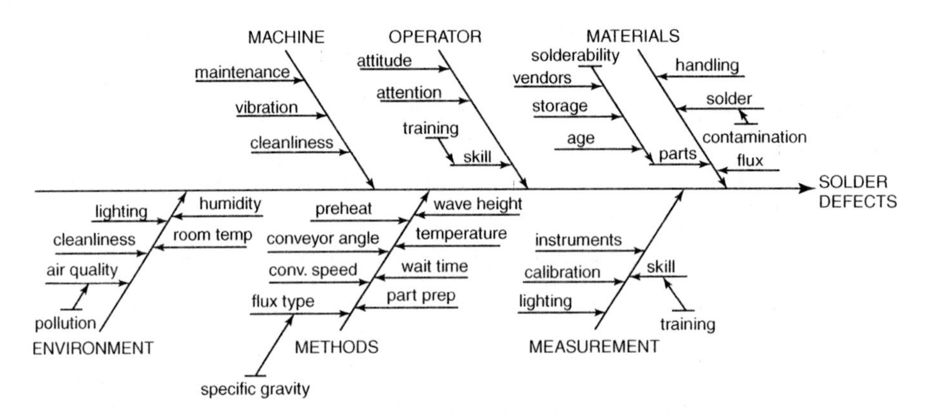

FIGURE 5.6 Completed cause-and-effect (fishbone) diagram.

defects: the parts themselves, handling of the materials, and the solder and flux used in the process. The chart points out that contamination can affect the solder's performance and also that the big issue affecting the parts is solderability. In the case of solderability, there are three levels of branches on the rib. Note that solderability can be affected by the vendor supplying the parts, storage of the parts before use, and age of the parts.

One could say, "The diagram didn't configure itself. Someone had to know the relationships before the diagram was drawn, so why is the diagram needed?" First, picture these relationships in your mind—no diagram, just a mental image. If you are not familiar with the process used in the example, pick any process involving more than two or three people and some equipment, such as the process of an athletic event. You will probably find it virtually impossible to be conscious of all the factors that come into play, much less how they relate and interact. Certainly, the necessary knowledge and information existed before the 35 factors were arranged in the cause-and-effect diagram. The key to the diagram's usefulness is that most likely no one individual had all that knowledge and information. This is why cause-and-effect diagrams normally are created by teams of people with divergent areas of expertise.

The team's initial effort is the development of the list of possible factors. This is usually done using brainstorming techniques. Such a list can be generated in a surprisingly short amount of time, usually no more than an hour. It is not necessary that the list be complete or even that all the factors be truly germane. Missing elements usually become obvious as the diagram is developed, and superfluous elements become evident as well. After the list has been assembled, all the team members contribute their knowledge and expertise in constructing the cause-and-effect diagram.

The completed diagram reveals factors or relationships that previously had not been obvious. The causes most responsible for the problem are usually isolated. Further, the diagram may suggest directions for action. It is possible in the example, for instance, that the team, because it is familiar with the plant's operation, could say with some assurance that solderability was an issue because of the parts being stored for long periods of time. The team might recommend switching to a just-in-time system, so that both storage and aging could be eliminated as factors affecting solderability. To bring performance up to best-in-class level, the problems isolated by the fishbone diagram will have to be corrected.

CHECK SHEETS

The *check sheet* is a third traditional performance-enhancement tool. The fuel that powers all these tools is data. In many companies, elaborate systems of people, machines, and procedures exist for the sole purpose of collecting data. This quest for data can become zealous to the point of obscuring the reason for data collection in the first place. Many organizations are drowning in data but thirsting for knowledge; they are "data rich and information poor." With the advent of powerful desktop computers, information collection has become an end unto itself in many instances.

Having access to data is essential. However, problems arise when trivial data cannot be winnowed from the important and when there is such a quantity of data that it cannot easily be translated into useful information. Check sheets help resolve these problems.

The check sheet is valuable in a wide variety of applications—its utility restricted only by the imagination of the person seeking information. The check sheet can take any form. The only rules are that (1) it must be designed so that data are entered using check marks and (2) displayed data must be easily translated into useful information. For example, it may take the form of a drawing of a product, with check marks entered at appropriate places to illustrate the location and type of finish blemishes. Or an accounts receivable department might create a check sheet to record the types and number of mistakes on invoices. Check sheets can apply to any work environment—not just to engineering, manufacturing, and construction.

The purposes of the check sheet are to make it easy to collect data for specific purposes and to present the data in a way that facilitates conversion to useful information. For example, suppose a company manufactures parts with a specified dimensional tolerance of 1.120″ to 1.130″. During the week, each part is measured and the data are recorded. Figure 5.7 is a table showing the week's results. (Note that this figure is not a check sheet.)

This figure contains all the data on shaft length for the week of July 11. Without a lot of additional work, it will be difficult to glean much useful information from this list. Imagine how much more difficult it would be yet if, instead of a table, you were presented with a stack of computer runs several inches thick. This is frequently the case in the information age. (The information age should be called the "data age," in our opinion, to reflect the difference between an abundance of raw, often meaningless data, and the paucity of truly useful information.)

A computer could be programmed to do something with this data to make it more useful, and in some situations that would be appropriate. After all, computers are good at digesting raw data and formatting it for human consumption. But before the computer can do that, some human must tell it exactly what to do, how to format the information, what to discard, what to use, and so on. If a programmer cannot first figure out what to do with the data, no amount of computer power will help. If, however, the company did know what it wanted to do with the data, the data could be preformatted so as to be instantly useful as it was being collected. This is one of the powerful capabilities of the check sheet.

The data in Figure 5.7 reports how the work being produced relates to the shaft length specification. The machine is set up to produce shafts in the center of the range so that normal variation does not spill outside the

Shaft Length—week of 7/11 (Spec 1.120 to 1.130)

Date	Length	Date	Length	Date	Length	Rom
11	1.125	11	1.128	11	1.123	
11	1.126	11	1.126	11	1.125	
11	1.119	11	1.123	11	1.122	
11	1.120	11	1.122	11	1.123	
12	1.124	12	1.126	12	1.125	
12	1.125	12	1.127	12	1.125	
12	1.121	12	1.124	12	1.125	
12	1.128	12	1.124	12	1.127	
13	1.123	13	1.125	13	1.121	
13	1.120	13	1.122	13	1.118	
13	1.124	13	1.123	13	1.125	
13	1.126	13	1.123	13	1.124	
14	1.125	14	1.127	14	1.124	
14	1.128	14	1.128	14	1.125	
14	1.126	14	1.123	14	1.124	
14	1.122	14	1.124	14	1.122	
15	1.124	15	1.121	15	1.123	
15	1.124	15	1.127	15	1.123	
15	1.124	15	1.122	15	1.122	
15	1.123	15	1.122	15	1.121	

FIGURE 5.7 Weekly summary of chart dimensional tolerance results. (This is not a check sheet.)

specified limits of 1.120″ and 1.130″ and thereby create waste. If the raw data could provide a feel for this as it is being collected, that would be very helpful. It would also be helpful to know when the limits are exceeded.

The check sheet in Figure 5.8 is designed to accept the data easily and at the same time to convert it to and display useful information. The check sheet actually produces a histogram of the data (see the following section for information about histograms). Data are obtained by measuring the shafts, as was done for Figure 5.7. But rather than logging the measured data by date, as was done in Figure 5.7, the check sheet in Figure 5.8 requires only noting the date (day of month) opposite the appropriate shaft dimension. The day-of-month notation serves as a check mark and at the same time keeps track of the day the reading was taken.

This check sheet should be set up on an easel on the shop floor, with entries handwritten in. The performance of the machine will then be visible at all times to operators, supervisors, engineers, or anyone else in the work area.

The data in Figure 5.7 are the same as the data in Figure 5.8. Figure 5.7 shows columns of sterile data. It will take hard work on someone's part to

Check Sheet

Shaft Length—Week of 7/11 (Spec: 1.120 to 1.130)

**Out of Limits

1.118**	13											
1.119**	11											
1.120	11	13										
1.121	12	13	15	15								
1.122	11	11	13	14	14	15	15	16				
1.123	11	11	11	13	13	13	14	15	15	15		
1.124	11	12	12	12	13	13	14	14	14	15	15	15
1.125	11	12	12	12	12	13	13	14	14			
1.126	11	12	12	13	14	14						
1.127	12	12	14	15								
1.128	11	11										
1.129	14											
1.130												
1.131**												
1.132**												

Enter day of month for data point

FIGURE 5.8 Check sheet of shaft dimensional tolerance results.

extract meaning from them. Assuming the data get translated into meaningful information, it probably will still remain invisible to the people who could make the best use of it—the operators. That can, of course, be overcome by more hard work, but in most cases the data will languish. On the other hand, Figure 5.8 provides a simple check sheet into which the data can be entered more easily, and once entered provide a graphical presentation of performance. Should the check sheet reveal that the machine is creeping away from the center of the range, or if the histogram shape distorts, the operator can react immediately. No additional work is required to translate the data into useful information, and no additional work is required to broadcast the information to all who can use it.

To set up a check sheet, think about your objectives. In this example, which involved making shafts to a specification, the company needed information about how well the machine was performing, a graphical warning whenever the machine started to deviate, and information about defects. Set up as a histogram, the check sheet met all these objectives. It is called a "process distribution check sheet" because it is concerned with the variability of a process. Other commonly used check sheets include defective item check sheets (detailing the variety of defects), defect location check sheets (showing where defects occur on the subject product), and defect factor check sheets (illustrating the factors—time, temperature, machine, or operator—that might influence defect generation). All check sheets do the same thing: They help identify root causes of performance problems so that the problems can be overcome and the relevant process(es) can be improved to the benchmarked level or better.

HISTOGRAMS

Histograms are used to chart frequency of occurrence. Any discussion of histograms must begin with an understanding of the two kinds of data commonly associated with processes: *attributes data* and *variables data*. Although they were not introduced as such, both kinds of data have been used in this chapter's examples. An attribute is a characteristic of a product or a process. The product has the attribute or it does not (see Figure 5.9). The example about making shafts of a specified length (see Figures 5.7 and 5.8) was concerned with measured data. That example involved shaft length measured in thousandths of an inch, but any scale of measurement can be used, depending upon the process under scrutiny. A process used in making electrical resistors would use electrical resistance in ohms, another process might use a weight scale, and so on. Variables data result from measurement.

Using the shaft example again, a common scenario in manufacturing plants would have been to place a Go-No-Go screen at the end of the process, which would accept all shafts between the specification limits of 1.120″ and 1.130″ and discard the rest. Data might have been recorded to track the number of shafts that were scrapped. Such a record might have looked like Figure 5.10, based on the original data.

Figure 5.10 tells us what we want to know if we are interested only in the number of shafts accepted versus the number rejected. Looking at the

Attributes Data versus Variables Data

Attributes Data

- Has or has not
- Good or bad
- Pass or fail
- Accept or reject
- Conforming or nonconforming

Variables Data

- Measured values (dimension, weight, voltage, surface, etc.)

FIGURE 5.9 The distinctions between attributes data and variables data should be understood.

Shaft Acceptance—Week of 7/11 (Spec: 1.120 to 1.130″)		
Date	**Accepted**	**Rejected**
11	11	1
12	12	0
13	11	1
14	12	0
15	12	0
Totals:	58	2

FIGURE 5.10 Summary data: weekly shaft acceptance.

shaft process in this way, we are using attributes data: The shafts either passed or failed the screening. This reveals only that between 3 and 4 percent of all the shafts were scrapped. It does not reveal anything about the process that may be contributing to the scrap rate. Nor does it tell us anything about how robust the process is. Might some slight change push the process over the edge? For that kind of insight, we need variables data.

One can gain much more information about a process when variables data are available. The check sheet in Figure 5.8 shows that both of the out-of-limits shafts were on the low side of the specified tolerance. The peak of the histogram seems to occur between 1.123″ and 1.124″. If the machine were adjusted to bring the peak up to 1.125″, some of the low-end rejects might be eliminated without causing any new rejects at the top end. The frequency distribution also suggests that the process as it now stands will always produce occasional rejects—probably in the 2 to 3 percent range at best.

SCATTER DIAGRAMS

The fifth of the performance-enhancement tools presented in this chapter is the *scatter diagram*. It is the simplest of the tools and one of the most useful. The scatter diagram is used to determine the correlation (relationship) between two characteristics (variables). Suppose you have an idea that there is a relationship between automobile fuel consumption and the rate of speed at which people drive. To prove or disprove such an assumption, you would record data on a scatter diagram: miles per gallon (mpg) on the *y*-axis, and miles per hour (mph) on the *x*-axis.

The scatter diagram in Figure 5.11 shows that the aggregate of data points contains a downward slope to the right. The correlation this demonstrates supports the thesis that the faster cars travel, the more fuel

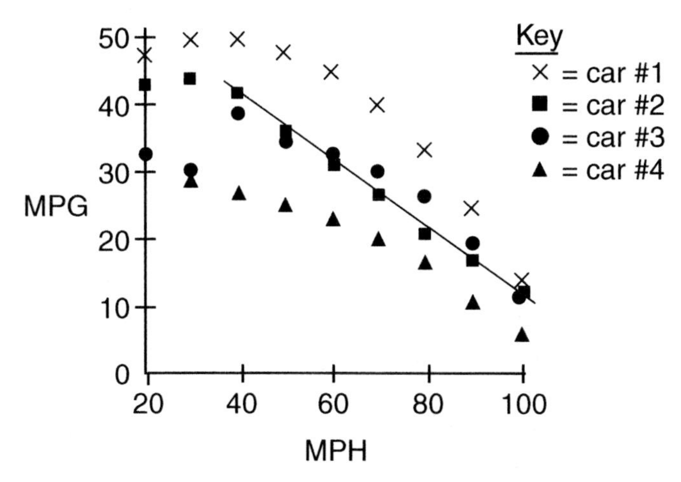

FIGURE 5.11 Scatter diagram: Speed versus fuel consumption for four automobiles.

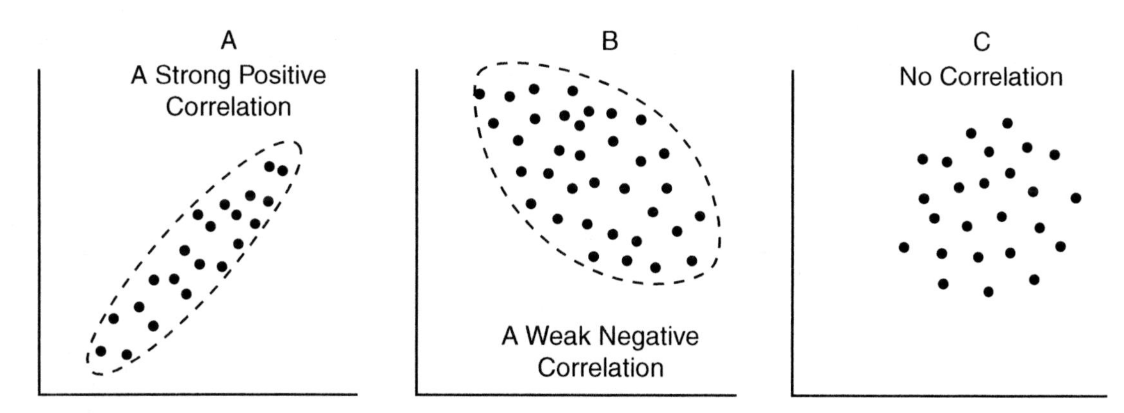

FIGURE 5.12 Scatter diagrams showing various correlations.

they use. Had the slope been upward to the right, as it actually appears to be (for three of the four cars) between 20 and 30 mph, then the correlation would have suggested that the faster you travel, the better the fuel mileage. Suppose, however, that the data points did not form any recognizable linear or elliptical pattern, but were simply a disorganized configuration. This would suggest that there is no correlation between speed and fuel consumption.

Figure 5.12 is a collection of scatter diagrams illustrating strong *positive correlation* (diagram A), weak *negative correlation* (diagram B), and *no correlation* (diagram C). To be classified as a strong correlation, the data points must be tightly grouped in a linear pattern. The more loosely grouped, the less the correlation—therefore the term "weak correlation." When a pattern has no discernible linear component, it is said to show no correlation.

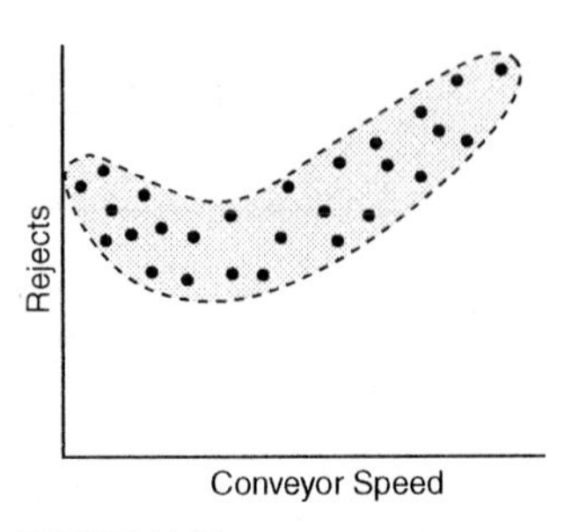

FIGURE 5.13 Scatter diagram: Conveyor speed versus rejects.

Scatter diagrams are useful in testing the correlation between process factors and characteristics of the product flowing out of the process. Suppose you want to know whether conveyor speed has an effect on solder quality in a machine soldering process. You could set up a scatter diagram with conveyor speed on the *x*-axis and solder rejects or nonconformities on the *y*-axis. By plotting sample data as the conveyor speed is adjusted, you can construct a scatter diagram to determine whether a correlation exists.

In this case, Figure 5.13 suggests that the correlation is a curve, with rejects dropping off as speed is initially raised, but then increasing again as the conveyor speed continues to increase. This is not atypical of process factors with optimum operating points. In the case of the conveyor, moving too slowly allows excess heat to build, causing defects. Increasing speed naturally produces better results, until the speed increases to the point where insufficient preheating increases the number of defects. Figure 5.13 not only reveals a correlation, but it also suggests that there is an optimum conveyor speed above or below which increased product defects will occur. Scatter diagrams are valuable performance-enhancement tools because they help prevent wasting of time on improvement strategies that will have little or no positive effects.

RUN CHARTS AND CONTROL CHARTS

These two types of charts—one, the run chart, more straightforward, and the other, the control chart, a more sophisticated outgrowth of the first—are usually thought of as a single tool. Taken together, they can be effective tools for tracking and controlling processes, and they are fundamental to process improvement.

Run Charts

A *run chart* records the performance of a process over time. The concept is strikingly simple, and indeed it has been used in modern times to track the performance of everything from AAA membership to zwieback production. Because one axis (usually the *x*-axis) represents time, the run chart can provide an easily understandable picture of what is happening in a process as time goes by. It causes trends to "jump out." For this reason, the run chart is also referred to as a "trend chart."

Consider as an example a run chart set up to track the percentage or proportion of product that is defective for a process that makes ballpoint pens. These are inexpensive pens, so production costs must be held to a minimum. On the other hand, many competitors would like to capture your share of the market, so at a minimum, you must deliver pens that meet customer expectations. A sampling system is set up that requires that a percentage of the process output be inspected. From each lot of 1,000 pens, 50 are to be inspected. If more than 1 pen from each sample of 50 is defective, the whole lot of 1,000 will be inspected. In addition to scrapping the defective pens, the objective is to discover why the defects were there in the first place and to eliminate the cause. Data from the sample will be plotted on a run chart. Because you anticipate improvements to the process as a result of this effort, the run chart will be an ideal indicator of success.

The run chart in Figure 5.14 is the result of sample data taken during a month—21 working days. The graph clearly shows that significant improvement in pen quality occurred during that time. Even better, the trend across the month was toward better quality (fewer defects). The most significant improvements came at the twelfth day and the seventeenth day, as causes for defects were found and corrected.

The chart can be continued indefinitely to track performance. Is it improving, staying the same, or losing ground? Scales may have to be changed, for clarity. For example, if you consistently found all samples had defects below 2 percent, it would make sense to change the *y*-axis scale to 0 to 2 percent. Longer-term charts would need to be changed from daily to weekly or even monthly plots.

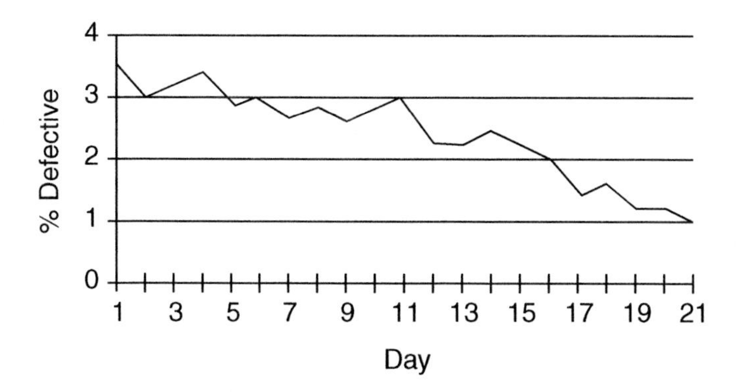

FIGURE 5.14 Run chart: Pen defect rate for a month.

The chart shows positive results during the first month of the pen-manufacturing process. What cannot be determined from the run chart, however, is what *should* be achieved. Assuming the defect rate can hold at two defective pens in 100, there are still 20,000 defective pens out of a million. Because only 5 percent of the pens produced are being sampled, you can assume that 19,000 of these find their way into the hands of customers—the very customers the competition wants to take away from you. So it is important to improve further. The run chart helps, but a more powerful tool is still needed.

Control Charts

The problem with the run chart is that it does not help users understand whether the variation is the result of *special causes*—such as changes in the materials used, machine problems, or lack of employee training—or *common causes,* which are purely random. Not until Dr. Walter Shewhart made that distinction in the 1920s was there a real possibility of improving processes using statistical techniques. Shewhart, then an employee of Bell Laboratories, developed the control chart to separate special causes from common causes.

Figure 5.15 shows a typical control chart. Data are plotted over time, just as with a run chart; the difference is that the data stay between the upper control limit (UCL) and the lower control limit (LCL) while varying about the center line or average *only so long as the variation is the result of common causes* (*i.e.*, statistical variation). Whenever a special cause (non-statistical cause) affects the process, one of two things happens: Either a plot point penetrates the UCL or LCL, or there is a "run" of several points in a row above or below the average line. When a penetration or lengthy run appears, this indicates that something is wrong that requires immediate attention.

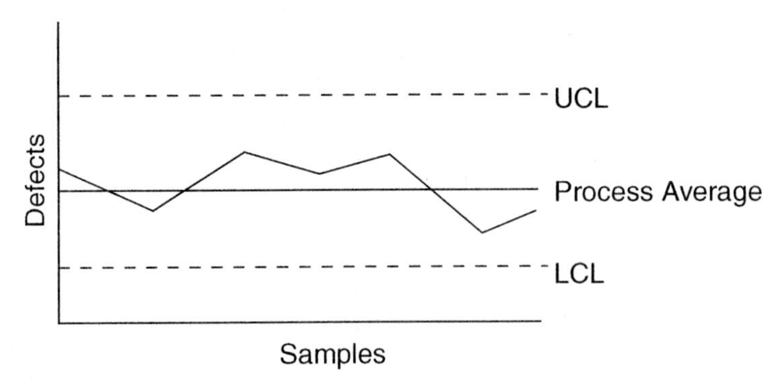

FIGURE 5.15 Basic control chart.

As long as the plots stay between the limits and don't congregate on one side or the other of the process average line, the process is in statistical control. If either of these conditions is not met, we say that the process is not in statistical control, or is "out of control"—hence the name "control chart."

Once you understand that it is the UCL, LCL, and process average lines added to the run chart that characterize a control chart, you may wonder how those lines are set. The positioning of the lines cannot be arbitrary, nor can the lines merely reflect how you want the process to perform. Such an approach does not help separate common causes from special causes, and it only complicates attempts at process improvement. UCL, LCL, and process average must be determined by valid statistical means.

All processes have built-in variability. A process that is in statistical control is affected by its natural random variability and exhibits the normal distribution of the bell curve. The more finely tuned the process, the less deviation there is from the process average and the narrower the bell curve. This is what makes it possible to define the limits and process average.

Control charts are appropriate tools for monitoring processes. A properly used control chart immediately alerts the operator to any change in the process being tracked. The appropriate response to the alert is to stop the process at once, preventing the production of defective products. Only after the special cause of the problem has been identified and corrected should the process be restarted. Once the root cause is corrected, that problem should never recur. (Anything less, however, and it is sure to return eventually.)

Control charts also enable continuous improvement of processes. When a change is introduced to a process that is operated under statistical process control (SPC) charts, the effect of the change will be seen immediately. You can tell when an improvement has been made, and also when the change is ineffective or even detrimental. This validation of improvements allows you to retain those that are effective. Continuous improvement is enormously difficult when the process is not in statistical control, because the process instability masks the results, good or bad, of any deliberate changes.

OTHER HELPFUL TOOLS

The performance-enhancement tools presented in the preceding sections are those that have been found most useful for the broadest spectrum of users. We recommend three additional tools to complete the kit of any business enterprise: the flow chart, the survey, and the design of experiments (DOX).

Flowcharts

A flowchart is a graphic representation of a process. It enables all parties involved to begin with the same understanding of the process. It may be revealing at the start of the flowcharting exercise to ask several team members to flowchart the process independently. If their charts differ, a problem may be revealed at the outset. Another strategy would be to ask team members to chart how the process actually works and then chart how they think it should work. Comparing the two versions may identify causes of problems and suggest improvements. The most common flowcharting method is to have the team—which is made up of the employees who work with the process and those who provide input to the process—develop the chart. To be effective, the completed flowchart must accurately reflect the way the process actually works, not how it should work. In studying the flowchart, the team can determine what aspects of the process are problematic and where improvements can be made.

You may already be familiar with the flowchart, at least to the point of recognizing one when you see it. This tool has been used for many years in many ways. Here, its relevant application is for illustrating the inputs, steps, functions, and outflows of a process.

A standard set of flowcharting symbols is used internationally and can be applied to any process. The most commonly used symbols are shown in Figure 5.16. To illustrate their use, a simple flow diagram is shown in Figure 5.17. Flow diagrams may be as simple or as complex as is necessary. Our purpose for Figure 5.17 is to chart the major process steps for receiving and repairing a defective unit from a customer, so we did not include subprocess detail. If an intent of the flowchart was to provide information on the troubleshooting process, each troubleshooting step would be included. The rectangle labeled "Troubleshoot" represents an entire subprocess that itself could be expanded into a complex flowchart.

From the high-level flowchart shown in Figure 5.17, one can observe the following.

1. The customer's defective unit is received.
2. The problem is located and corrected.
3. The repaired unit is tested.
4. If the unit fails the test, it is recycled through the repair process until it does pass.
5. After the unit passes the test, paperwork is completed.
6. The customer is notified.
7. The unit is returned to the customer along with a bill for services.

With this flowchart as a guide, the next step is to develop detailed flowcharts of the subprocesses you want to improve. Only then can you understand what is really happening inside the process, which steps add value,

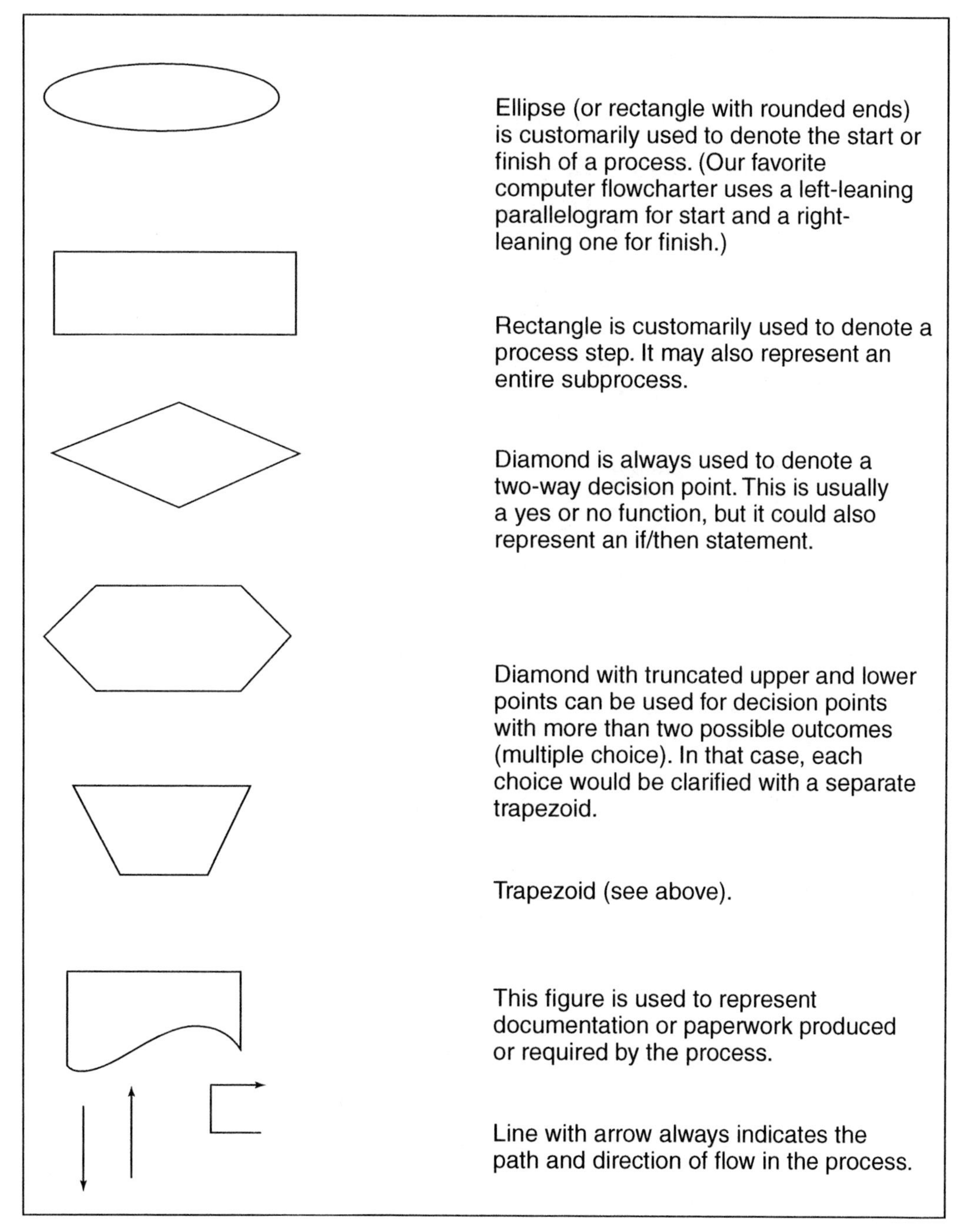

FIGURE 5.16 These common flowcharting symbols fit virtually any situation.

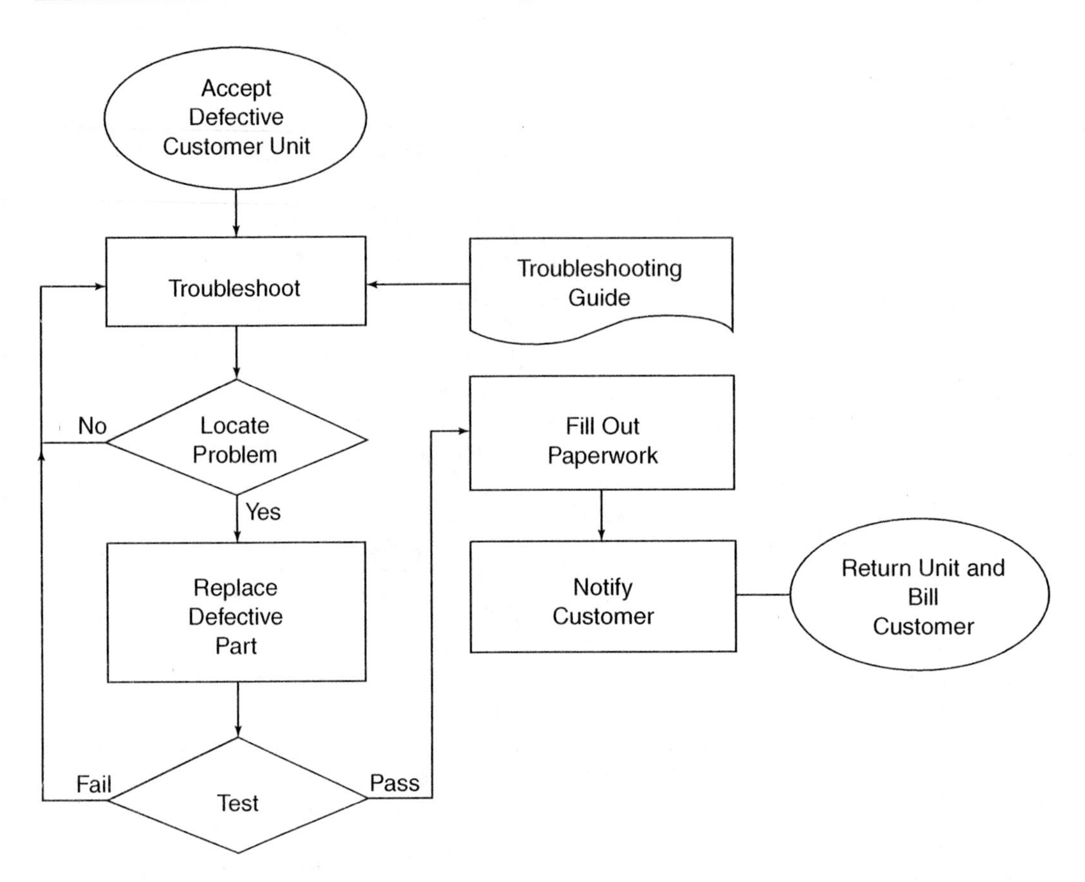

FIGURE 5.17 Typical process flowchart.

where time is being consumed, where there are redundancies, and so on. Once the process has been flowcharted, it is almost always easy to see potential areas for improvement and streamlining. Without the flowchart, it may be impossible.

More often than not, people who work directly with a process are amazed to find how little they understood the process before they had seen it flowcharted. Working with any process day in and day out tends to breed a false sense of familiarity.

The authors once worked with a large manufacturing operation that was having a major problem with on-time delivery of systems worth \$500,000 to \$2,000,000. There were lots of reasons for the difficulty, but a fundamental problem was that the company was not getting the input materials on time—even with a 24-month lead time for delivery. One of the first things we did was flowchart the entire material system. We started the chart at the signing of the customer's order and completed it at the point where the material was delivered to the stockroom. The chart showed dozens of people involved, endless loops for approval and checking, and flawed subprocesses that consumed enormous amounts of time.

When the flowchart was finished, it was clear that, with the current process, the best case from the start of the order cycle until material could be expected to be in hand was 55 weeks. The worst case could be easily double that. With this knowledge, we attacked the material process and quickly whittled it down to 16 weeks, and from there to 12 weeks.

Here was a process that had grown over the years to the point that it was no longer efficient. But the individual players in the process didn't see the problem. They were working very hard, doing what the process demanded and fighting the fires that constantly erupted when needed material was not available. The flow diagram illuminated the problems with the process and showed what needed to be done.

If you set out to control or improve any process, it is essential that you fully understand it and the reasons for it. Don't make the assumption that you already know, or that the people working with the process know, because chances are that you don't and they don't. Work with the people who are directly involved, and flowchart the process as a first step in the journey to world-class performance. Not only will you better understand how the process works, but you will spot unnecessary functions or weaknesses and be able to establish logical points in the process for control chart application. The flowchart will suggest the use of other tools, as well.

Surveys

At first mention, the survey may not seem useful. When you think about it, though, all performance-enhancement tools are designed to present information—valuable information that is pertinent and easily understood. The purpose of a survey is to obtain relevant information from sources that otherwise would not be heard from—at least not in the context of providing helpful data. Because you design your own survey, you can tailor it to your needs. We believe that the survey meets the criteria for being a performance-enhancement tool. Experience has shown that the survey can be very useful.

Surveys can be conducted with "internal customers," employees who provide feedback on problem areas relating to products or services. They can also be conducted with external customers—your business customers—to learn how your products or services rate in their eyes. The customer (internal or external) orientation of the survey is important, because the customer is the authority on the quality of your goods and services that matters most. Some companies conduct annual customer satisfaction surveys. They use the input from these surveys to focus their improvement efforts.

Surveys increasingly are being used with suppliers as well. Companies are finally realizing that having a huge supplier base is not the good thing they once thought it was. The tendency today is to cut back drastically on the number of suppliers, retaining only those that offer the best value and that are willing to enter into partnership arrangements. If a company goes this route, it had better know how satisfied the suppliers are with the working relationship and what they think of future prospects. The survey is a good tool for determining this. It is possibly the best method for starting a supplier reduction/supplier partnership program.

Even if you are not planning to eliminate suppliers, it is vital that you know what your suppliers are doing. It would make little sense to go to the trouble of implementing ECS if your suppliers continue doing business as usual. As you improve your processes, services, and products, you cannot afford to be hamstrung by poor quality from your suppliers. Surveys are the least expensive way of determining where suppliers stand on ECS principles and what their plans are for the future. The survey can also be a not-too-subtle message to the suppliers that they had better get with the program.

A typical department in any organization has both internal suppliers and internal customers. A survey with a customer-oriented point of view has proven to be a powerful tool for opening communication among departments and getting them to work together toward companywide goals rather than just for department agendas.

Design of Experiments

Design of experiments (DOX) is a sophisticated method of experimenting with processes for the purpose of optimizing them. If you deal with complicated processes, DOX may be the only practical way to bring about improvement. Such a process might be illustrated by a wave soldering machine. There are many wave solder process factors, including:

- conveyor speed
- conveyor angle
- flux specific gravity
- flux type
- solder type
- solder temperature
- preheat temperature
- PC board layer count
- PC board groundplane mass
- wave height

These 10 factors influence the process, often interacting with one another. The traditional way of determining the proper selection/setting has been to vary one factor while holding all others fixed. That kind of experimentation leads to making hundreds of individual runs for even the simplest process. It is unusual to arrive at the optimum setup, because a change in one factor frequently requires adjustment of one or more of the other factors for best results.

Design of experiments reduces the number of runs from hundreds to tens as a rule, or by an order of magnitude. This means of process experimentation allows simultaneous multiple-factor adjustment, which shortens the total process, but equally as important, reveals complex interaction among the factors. A well-designed experiment can be conducted on a process like wave soldering in 30 to 40 runs and can establish the optimum setting for each adjustable parameter for each of the selected factors. For example, optimal settings for conveyor speed, conveyor angle, wave height, preheat temperature, solder temperature, and flux specific gravity can be established for each PC board type and solder alloy.

DOX also shows which factors are critical and which are not. This information enables you to set up control charts for those factors that matter and to avoid expending effort on the ones that don't. While design of experiments is beyond the scope of this book, the DOX work of Deming, Taguchi, and others may be of help to you. Remember that DOX is an available tool when you start trying to improve a complex process.

RATIONALE FOR CONTINUAL IMPROVEMENT

Continual improvement is fundamental to success in the global marketplace. Companies that maintain the status quo in customer satisfaction are like a runner who stands still in a race. Competing in the global marketplace is like competing in the Olympics: Last year's records are sure to be broken this year. Athletes who don't improve continually are not likely to remain in the winner's circle for long. The same is true of companies that compete globally.

Customer needs are not static. They change continually. A product feature that is considered innovative today will be considered routine tomorrow. A price that is considered a bargain today will be too high to be competitive tomorrow. A good example in this regard is the ever-falling price of personal computers, even though new features are added. The only way a company can hope to compete in the modern marketplace is to continually improve.

ESSENTIAL IMPROVEMENT ACTIVITIES

Continual improvement is not about solving isolated problems as they occur. Such an approach puts out fires rather than preventing them. Solving a problem without correcting the fault that caused it means the problem will occur again. Quality expert Peter R. Scholtes recommends the following activities as being crucial to continual improvement.[1]

- *Maintain communication.* Communication is essential to continual improvement. This cannot be overemphasized. Communication within improvement teams and between teams is a must. It is important that information be shared before, during, and after attempting to make improvements. All people involved, as well as any person or unit affected by a planned improvement, should know what is being done and why.
- *Correct obvious problems.* Most process problems are not obvious, and a great deal of study is required to isolate them and find solutions. This is

why the scientific approach is so important. However, there are times when a process problem is obvious. In such cases, the problem should be corrected immediately. Days spent studying a problem for which the solution is obvious for the sake of using a scientific approach can result in $100 solutions to $10 problems.

- *Look upstream.* Look for causes, not symptoms. This is a difficult point to make with people who are used to taking a cursory glance at a situation and fixing it without taking the time to determine what caused it.

- *Document problems and progress.* Take the time to write it down. It is not uncommon for an organization to solve the same problem over and over because nobody took the time to document the problem and its previous solutions. The fundamental rule is: "Document, document, document."

- *Monitor changes.* Regardless of how well a problem is studied, the solution may not solve it or may only partially solve it. For this reason, it is important to monitor process performance after changes have been implemented. It is also important that pride of ownership on the part of those who recommended the change not interfere with objective monitoring. Monitoring activities are essential regardless of how the improvement effort is structured.

STRUCTURE FOR IMPROVEMENT

Improvement doesn't happen by chance. It must be undertaken in a systematic, step-by-step manner. If an organization is to make continual improvements, it must be structured appropriately.

- *Establish a quality-improvement council.* The quality-improvement council in an organization has overall responsibility for continual improvement. Its responsibility is to implement, monitor, and institutionalize improvements. It is essential that the membership include executive-level decision makers.

- *Develop a statement of responsibilities.* It is essential that all members of the quality-improvement council, as well as employees who are not currently members, understand the council's responsibilities. One of the first priorities of the council is to develop and distribute a statement of responsibilities, signed by the organization's CEO. Responsibilities that should be stated include the following: (1) formulating policy as it relates to ECS; (2) setting benchmarks and dimensions (cost of poor customer service); (3) establishing the team and project-selection processes; (4) providing the necessary resources (training, time away from job duties, and so on); (5) implementing the project; (6) establishing measures for tracking progress and undertaking monitoring efforts; and (7) implementing an appropriate reward and recognition program.

- *Establish the necessary infrastructure.* The quality-improvement council constitutes the nucleus of an organization's ECS effort. There is more to the ECS infrastructure than just the council, however. The remainder of the infrastructure consists of council subcommittees assigned responsibility for specific duties, project-improvement teams, quality-improvement managers, an ECS training program, and a structured improvement process.

SCIENTIFIC APPROACH TO IMPROVEMENT

The scientific approach is one of the fundamental methodologies that separates the ECS approach from other ways of doing business. The scientific approach amounts to "making decisions based on data, looking for root causes of problems, and seeking permanent solutions instead of relying on quick fixes."[2]

Scholtes developed the following four strategies for putting the scientific approach to work.[3]

- *Collect meaningful data.* Meaningful data are free from errors of measurement or procedure and have direct application to the issue in question. It is not unusual for an organization or a unit within it to collect meaningless data or to make a procedural error that results in erroneous data. In fact, in the age of computers, this is quite common. Decisions based on meaningless or erroneous data are bound to lead to failure. Before collecting data, decide exactly what is needed, how it can best be collected, where the data exists, how it will be measured, and how to determine that the data are accurate.
- *Identify root causes of problems.* The strategy of identifying root causes is emphasized throughout this book. Too many resources are wasted attempting to solve symptoms rather than problems. The performance-enhancement tools presented earlier in this chapter are helpful in separating problems from causes.
- *Develop appropriate solutions.* With the scientific approach, solutions are not assumed. Collect the relevant data, make sure it is accurate, identify root causes, and then develop a solution that is appropriate. Too many people begin with "I know what the problem is. All we have to do to solve it is . . . " When the scientific approach is applied, the identified problem is often much different from what would have been suspected using a hunch or intuition. As a result, the solution is also different.
- *Plan and make changes.* Too many decision makers use the "ready-fire-aim" approach rather than engaging in careful, deliberate planning. What is out of sequence with that approach is planning (aim). People act first and plan later. Just as one should aim before shooting, one should plan before acting. Planning forces you to look ahead, anticipate needs and related resources, foresee problems, and consider how problems should be handled.

The scientific approach has to do with establishing reliable performance indicators and using them to measure actual performance. Giorgio Merli lists the following useful process performance indicators:[4]

- Number of errors or defects
- Number of repetitions of work tasks
- Efficiency indicators (e.g., units per hour, items per person)
- Number of delays
- Duration of a given procedure or activity
- Response time or cycle
- Usability/cost ratio
- Amount of overtime required
- Changes in workload
- Vulnerability of the system
- Level of criticalness
- Level of standardization
- Number of unfinished documents

This is not a complete list. Many other indicators could be added. Those actually used vary widely from organization to organization. Regardless of which are actually used, however, such indicators are an important aspect of the scientific approach.

DEVELOPMENT OF IMPROVEMENT PLANS

After a performance-improvement project has been selected, a project-improvement team is established. The team should consist of representatives from every unit that will be involved in carrying out improvement strategies. The project improvement team should begin by developing an improvement plan, to make sure the team does not take the "ready-fire-aim" approach mentioned earlier.

The first step is to develop a mission statement for the team. This statement should clearly define the team's purpose and should be approved by the company's quality-improvement council. After this has been accomplished, the plan can be developed. The plan is developed as explained in the following list.

1. *Understand the process.* Before attempting to improve a process, make sure every team member thoroughly understands it. How does it work (flowchart)? What is it supposed to do? What are the best practices known pertaining to the process (benchmarking)? The team should ask these questions and pursue the answers together. This will give team members a common understanding, eliminate ambiguity and inconsistencies, and

point out any obvious problems that must be dealt with before proceeding to the next stage.

2. *Eliminate errors.* In analyzing the process, the team may identify obvious errors that can be eliminated quickly. Such errors should be corrected before proceeding to the next stage. This stage is sometimes referred to as "error proofing" the process.

3. *Remove slack.* This stage involves analyzing all the steps in the process to determine whether they serve any purpose and, if so, what that purpose is. In any organization, people follow processes that have been in place for years, without wondering why things are done a certain way, if they could be done better another way, or if they need to be done at all. There are few processes that cannot be streamlined.

4. *Reduce variation.* Variation in process performance results either from common causes or from special causes. Common causes result in slight variations and are almost always present. Special causes typically result in greater variations and are not always present. Strategies for identifying and eliminating sources of variation are discussed in the next section of this chapter.

5. *Plan for continual improvement.* By the time this step has been reached, the process should be in good shape. The key now is to incorporate the improvements into the process permanently, so that continual improvement becomes a normal part of doing business. The Plan-Do-Check-Act cycle applies here. With this cycle, each time a problem or potential improvement is identified, an improvement plan is developed (plan), implemented (do), monitored (check), and refined as needed (act).

COMMON IMPROVEMENT STRATEGIES

Numerous and varied processes are used in engineering, manufacturing, and construction. Consequently, there is no single road map to follow for improving them. However, a number of standard strategies can be used as a menu from which appropriate process-improvement strategies can be selected. Figure 5.18 shows several standard strategies that can be used to improve processes on a continual basis. These strategies are explained in the following sections.[5]

Describe the Process

The strategy of describing the process is used to make sure that everyone involved in improving a process has a detailed knowledge of it. Usually this requires some investigation and study. The steps involved are as follows:

1. Establish boundaries for the process.
2. Flowchart the process.

Process Improvement Strategies

✓ Describe the process.
✓ Standardize the process.
✓ Eliminate errors in the process.
✓ Streamline the process.
✓ Eliminate variation.
✓ Get the process in statistical control.
✓ Improve the design.

FIGURE 5.18 Most processes can be improved significantly through the application of these strategies.

3. Make a diagram of how the work flows.
4. Verify your work.
5. Correct immediately any obvious problems that are identified.

Standardize the Process

To continually improve a process, all people involved in its operation must be using the same procedures. This often is not the case. It is important to ensure that *all* employees are aware of using the best, most effective, most efficient procedures known. The steps involved in standardizing a process are as follows:

1. Identify the currently known best practices and write them down (benchmarking).
2. Test the best practices to determine if they are, in fact, the best, and improve them if there is room for improvement. (These improved practices then become the new benchmarks.)
3. Make sure that the new improved process becomes the standardized process.
4. Keep records of process performance, update them regularly, and use them to identify ways to improve the process even further on a continual basis.

Eliminate Errors in the Process

This strategy involves identifying common errors in the operation of the process and then eliminating them. The goal is to avoid steps, procedures, and practices that are done a certain way simply because they have always been done that way.

Streamline the Process

The strategy of streamlining is used to remove slack from a process. This can be done by reducing inventory, reducing cycle times, and eliminating unnecessary steps. After a process has been streamlined, each step in it has significance, contributes to the desired end, and adds value.

Reduce Sources of Variation

The first step in the strategy of reducing sources of variation is identifying those sources. They can often be traced to differences among people, machines, measurement instruments, material, sources of material, operating conditions, and times of day. Differences among people can be attributed to levels of capability, training, education, experience, and motivation. Differences among machines can be attributed to age, design, and maintenance. Regardless of the source of variation, after it has been identified, the information should be used to reduce the amount of variation to the absolute minimum. For example, if the source of variation is a difference in the levels of training for various operators, training should be provided for those who need it. If one set of measurement instruments is not as finely calibrated as another, it should be brought up to equal calibration.

Bring the Process Under Statistical Control

A control chart is planned, data are collected and charted, special causes are eliminated, and a plan for continual improvement is developed.

Improve the Design of the Process

There are many ways to design and lay out a process, most of which can be improved on. The best way to improve the design of a process is through an active program of experimentation. To produce the best results, an experiment must be properly designed. This involves the following steps:

1. Define the objectives of the experiment. What factors do you want to improve? What specifically do you want to learn from the experiment?
2. Decide which factors are going to be measured (*e.g.*, cycle time, yield, or finish).
3. Design an experiment that will measure the critical factors and answer the relevant questions.

4. Set up the experiment.
5. Conduct the experiment.
6. Analyze the results of the experiment.
7. Act on the results of the experiment.

Summary

1. Pareto charts are useful for separating the important from the trivial. They are named after Italian economist/sociologist Vilfredo Pareto, who developed the theory that a minority of causes are responsible for a majority of problems. Pareto charts are important because they help an organization decide where to focus its limited resources. On a Pareto chart, data are arrayed along an *x*-axis and a *y*-axis.

2. The cause-and-effect diagram was developed by the late Dr. Kaoru Ishikawa, a noted Japanese quality expert; it is sometimes called the Ishikawa diagram. Its purpose is to identify and isolate the causes of problems. It is the only one of the seven basic quality tools that is not based on statistics.

3. The check sheet is a tool that helps separate unimportant data from important information. Check sheets make it easy to collect data for specific purposes and to present it in a way that facilitates its conversion into useful information.

4. Histograms have to do with variability. There are two kinds of data commonly associated with processes: attributes data and variables data. An attribute is a characteristic of a product or a process; the product has the attribute or it does not. Variables data result when something is measured. A histogram has a measurement scale on one axis and a frequency of like measurements on the other.

5. The scatter diagram is the simplest of the seven basic quality tools. It is used to determine the correlation between two variables. It can show a positive correlation, a negative correlation, or no correlation.

6. Run charts and control charts are typically used together as one tool. The control chart is a more sophisticated version of the run chart. The run chart records the output results of a process over time. For this reason, the run chart is sometimes called a "trend chart." The weakness of the run chart is that it does not tell whether the variation results from special causes or common causes. This weakness gave rise to the control chart. On a control chart, data are plotted just as they are on a run chart, but a lower control limit (LCL), an upper control limit (UCL), and a process average (center line) are added. The plotted data stays between the UCL and the LCL while varying about the center line only so long as the variation is the result of common causes such as statistical variation.

7. Other useful quality tools are the flow diagram, the survey, and the design of experiments (DOX). Flowcharts are used in ECS for charting the inputs, steps, functions, and outflows of a process. From this, one can understand more fully how the function works, who or what influences the process, the inputs and outputs, and even the timing. The survey is used to obtain relevant information from sources that otherwise would not be heard from in the context of providing helpful data. DOX is a sophisticated method for experimenting with processes for the purpose of optimizing them.

8. The rationale for continual improvement is that it is necessary to compete in the global marketplace. Maintaining the status quo, even if the status quo is of high quality, is like standing still in a race.

9. Essential improvement activities include the following: maintaining communication, correcting obvious problems, looking upstream, documenting problems and progress, and monitoring change.

10. Structuring for quality improvement involves the following: establishing a quality-improvement council, developing a statement of responsibilities, and establishing the necessary infrastructure.

11. Using the scientific approach means collecting meaningful data, identifying root causes of problems, developing appropriate solutions, and planning and making changes.

12. Developing improvement plans involves the following steps: understanding the process, eliminating obvious errors, removing slack from the process, reducing variation in the process, and planning for continuous improvement.

13. Commonly used improvement strategies include the following: describing the process, standardizing the process, eliminating errors in the process, streamlining the process, reducing sources of variation, bringing the process under statistical control, and improving the design of the process.

Key Phrases and Concepts

Attributes data
Cause-and-effect diagram
Check sheet
Common causes
Communication
Continual improvement
Control chart
Design of experiments (DOX)
Flexible schedules
Flowchart
Flow production
Histogram
Improvement plans
In statistical control
Localize problems
Lower control limit
Obvious problems
Pareto chart
Pareto principle
Plan-Do-Check-Act

Process average
Process variability
Quality-improvement council
Run chart
Scatter diagram
Scientific approach
Special causes
Survey
Trend chart
Upper control limit
Variables data
Variation

Review Questions

1. Explain the purpose of a Pareto chart. Give an example of a situation in which a Pareto chart would be used.
2. Describe the origin and use of fishbone diagrams.
3. How would a check sheet be used in a modern production facility?
4. What is a histogram, and how is a histogram used?
5. What is the purpose of a scatter diagram? Give an example of how a scatter diagram would be used.
6. Contrast and compare run charts and control charts.
7. What purpose is served by flowcharts?
8. Give an example of how a survey might be used in a modem production setting.
9. What is the purpose of design of experiments (DOX)?
10. Explain the rationale for continuous improvement.
11. What are the five essential improvement activities?
12. If you were an executive manager in an organization, how would you structure the organization for quality improvement?
13. What is meant by using the scientific approach?
14. Describe the steps involved in developing an improvement plan.
15. List and explain three widely used improvement strategies.

ECS APPLIED: DIVERSIFIED TECHNOLOGIES COMPANY BEGINS ITS IMPROVEMENT PROJECTS

In the last installment of this case, two of DTC's vice presidents—Meg Stanfield (engineering) and Tim Wang (manufacturing)—gave CEO David Stanley an update on the benchmarking projects in their respective divisions. Both divisions had successfully benchmarked all their key processes, including engineering, finance, human resources, production, and quality. Stanfield had indicated that the work in her division would most need to focus on procurement, production control, and assembly. Stanley was pleased with the reports of these two vice presidents, but he was becoming increasingly concerned about the failure of DTC's vice president for construction, Conley Parrish, to get the ECS implementation started in his division.

David Stanley began the meeting of DTC's vice presidents on a somber note. "Tim and Meg, I want to hear your reports on improvement projects, but before we get into that I have an announcement to make." The two vice presidents could tell from the look on Stanley's face and the tone of his voice that they were about to get bad news. "Conley Parrish has given me his letter of resignation," said Stanley, clearly upset. "I've offered his position to Jake Arthur, Conley's number two. Jake has accepted the job and will join us at future meetings."

The room fell silent. For a few moments no one spoke. Conley Parrish was an effective vice president. His division, although not the top performing division at DTC, was solid and dependable. It always turned a profit. Stanfield and Wang had worked with Conley Parrish for years. They knew his wife and children; they had even helped him through the trauma of losing a teenage son in an automobile accident. Parrish had built both of their homes. His resignation would be felt throughout DTC, both professionally and personally. David Stanley was feeling it already. He and Conley Parrish had been friends since college.

Tim Wang finally broke the silence. "Why did he quit?" David Stanley looked as if the question pierced his heart. "He said he doesn't like where I am taking the company. It's the ECS implementation. Conley does not buy into the concept. He thinks the whole thing is a waste of time." Stanley went on to explain that Parrish planned to use his severance pay, which would be substantial, to start his own company. "He'll be a tough competitor," said Meg Stanfield. "We'll lose some customers," Tim Wang agreed. "I know," responded Stanley, shaking his head in disbelief at the turn of events.

"I've thought some about the customer side of the issue," said Stanley. "Conley will take some work from us at first, there is no doubt about that. But I don't think the migration will last. If Conley isn't willing to concern himself with customer service, he won't last long in this industry. Frankly, I'm more worried about his company failing than I am that it might succeed." Stanley went on to explain that he had already scheduled meetings with DTC's best construction customers. He planned to lay the situation out for them forthrightly, leaving nothing out. "When they see why Conley left us, and when they see that Jake is taking his place, most of them will probably give us a chance to keep their business."

With the Conley Parrish issue out of the way, Stanley listened while the two vice presidents updated him on the various improvement projects underway in their divisions. According to Meg Stanfield, improving the efficiency of the research and development function was going to be tough, but several obvious problems had already been solved. "David, you need to know that we might need to retire Marsha Atwood. Even after being with us all of these years, she still operates as if she is in an academic environment and receiving government funding. Her projects are always the slowest movers." "Why don't you get with your human resources folks and see what kind of exit package we can put together for Marsha," answered Stanley. "I know she would like to spend more time with her grandchildren anyway. She will probably welcome a retirement offer that is designed to meet her needs."

Tim Wang gave a presentation on his various improvement projects. "We are making the transition to statistical process control (SPC) right now. It's going to make a big difference in production control. On the procurement side, we have used Pareto charts to identify several problem suppliers. I think we will be going to a supplier-certification program in the near future." After hearing their reports, Stanley thanked his two vice presidents and asked them to look in on Jake Arthur in the near future to welcome him to the team. The resignation of Conley Parrish was a blow to the company and to Stanley personally, but DTC was moving in the right direction. "It's going to be tough in the construction division for a while," thought Stanley. "But we'll get through it, and in the long run we will be a better company."

DISCUSSION CASES

The following cases provide examples of how the various concepts presented in this chapter might play out in actual companies. The cases are provided to prompt discussion, give the reader a feel for the types of problems confronted in the workplace, and reinforce the ECS concept in question.

CASE 5.1 DHI Puts Its "Punchlist" on a Pareto Chart

Dream Homes, Inc. (DHI) is a residential contractor specializing in upscale homes in the southwestern United States. When one of its new homes is sold, DHI conducts a final inspection in conjunction with the buyer and develops a summary of defects to be corrected. This summary is called a "punchlist." Mack Turgood, DHI's construction superintendent, would like to reduce the length of these punchlists to the absolute minimum. This, he believes, will simultaneously decrease costs and improve customer satisfaction.

To get started, Turgood assembled data from the punchlists of the last 30 homes DHI sold. He found the defect types and occurrence rates indicated in the following table.

Defect Type	Occurrence
Damaged walls	13
Exterior paint	5
Plumbing	33
Caulking	28
Electrical	25
Cabinetry	12
Woodwork	46
Doors	14
HVAC	11
Roof	3
Masonry	2
Interior paint	61
Landscaping	16
Fixtures	7

With this data compiled, Turgood constructed a Pareto chart to view it graphically. The chart clearly showed that Turgood could make the most substantial improvements in the shortest time by focusing his immediate efforts on interior paint defects, plumbing problems, caulking defects, and woodworking discrepancies. He decided that exterior paint defects, roof problems, and masonry discrepancies did not occur with sufficient regularity to warrant immediate attention. He also decided that once he had dealt sufficiently with his list of primary concerns, he would then turn his attention to what he thought of as the secondary issues (damaged walls, cabinetry, doors, HVAC, and landscaping). Using his Pareto chart, Turgood had no trouble convincing higher management of the wisdom of his plan.

Discussion Questions

1. Have you ever used a Pareto chart to display data? Explain.
2. Ask 10 or more people the following question: "What is your favorite flavor of ice cream?" Display their responses on a Pareto chart.

CASE 5.2 Quantum Manufacturing and Engineering Stops Putting out Fires

Brenda Allison, CEO of Quantum Manufacturing and Engineering (QME), was sure she knew what had gone wrong, but just to be sure, she had hired a process-improvement consultant. That was two months ago. Today the consultant had given Allison his report. She had been right. Her company was not reaching its process benchmarks because key people were spending too much time putting out fires. "Ms. Allison, Quantum is never going to make much headway in process improvement until you get some basics in place," said the consultant. "What do you mean? What are the basics we're missing?"

"Let's begin with communication. Your engineering and manufacturing managers are trying to do everything themselves. They're not involving all the process stakeholders, and they're not communicating with everyone who should be involved. The first thing you need to do to improve a process is flowchart and document the process." "We did that," responded Allison. "No, your managers and supervisors did it. They didn't involve all the process stakeholders. Consequently, your flowcharts show how the processes are supposed to work, not how they actually do work. I recommend that you select just one key process, identify every stakeholder for the process, and involve them all in flowcharting that process. Once you have a realistic flowchart, then communicate clearly where the process's performance is at the moment and where it needs to be. Then, involve every stakeholder in making recommendations concerning how to improve the process. Get started making the improvements and communicate constantly." "That is a little different from the approach we've been using, isn't it?" said Allison with a touch of embarrassment.

The consultant went on to explain that obvious problems with a process should be corrected right away instead of being studied, and that Quantum's managers and employees needed to look for root causes rather than symptoms. He also encouraged Allison to have her process-improvement teams document all the problems they identified, steps taken to correct the problems, and progress made. Finally, the consultant explained that all process-improvement strategies should be monitored constantly. Not everything they tried would work.

"There is one final issue I need to caution you about," said the consultant. "It has to do with the issue of monitoring. It has been my experience that pride-of-ownership is the Achilles heel of process improvement. Someone will recommend an improvement strategy and invest so much pride in it that he or she refuses to acknowledge the obvious when it doesn't work. As the CEO, you have to ensure this doesn't happen. You don't want your employees taking blind shots in the dark. When someone suggests an improvement strategy, it should be thoroughly analyzed and widely discussed. However, once the stakeholders agree that the strategy has merit, it should be implemented without any risk of retribution should it fail to produce the desired results. This is critical."

Allison read the consultant's report from cover to cover twice. Then, when she was sure she knew what needed to be done, she gave copies of the report to all Quantum's management personnel. She gave them two days to study the report and then called a meeting to discuss it. Within just weeks, Quantum's process-improvement program was on track and making progress.

Discussion Questions

1. Have you ever known someone who insisted that his idea was good in spite of the facts just because it was his idea? How was this situation resolved?
2. What problem might occur if a process-improvement strategy is implemented but not monitored?

Endnotes

[1]Peter R. Scholtes, *The Team Handbook* (Madision, WI: Joiner Associates, 1992), pp. 5-6–5-9.

[2]Ibid, p. 5-9.

[3]Ibid.

[4]Giorgio Merli, *Total Manufacturing Management* (Cambridge, MA: Productivity Press, 1990), p. 143.

[5]Scholtes, *The Team Handbook*, pp. 5-54–5-67.

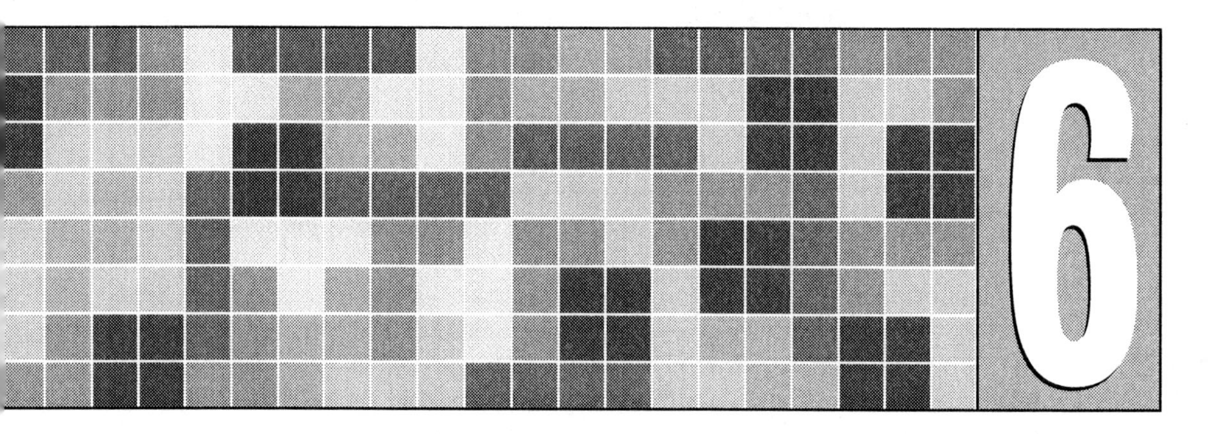

Provide Training for Employees and Customers

Often the difference between a mediocre athlete and a world-class athlete is training. This is also true of mediocre and world-class companies.

GOALS

- Define training.
- Describe the need for training.
- Discuss how to assess training needs.
- Demonstrate how to provide training.
- Explain how to evaluate training.
- Describe the role of supervisors, managers, and others in providing trainers.

- List the fundamental ECS training topics.
- Discuss the benefits to an organization of customer training.

Training is not initiated until Step 6 in a 10-step model. This may strike some as odd. Why not begin the implementation of ECS with training? This is a reasonable question, but there are sound answers. Training is most effective when delivered on a just-in-time basis. When training is delivered too early, employees do not have a sufficient appreciation of their need for it. Consequently, the necessary motivation may not be there. In addition, the skills developed in training must be applied right away, or they become stale—or even worse, they are lost. Providing training at the beginning of the ECS implementation would be too early, but by Step 6 enough activity has taken place to generate interest and a sense of need in employees. And by the time the company has reached Step 6, employees are able immediately to apply what they learn in this training. It should also be remembered that ECS is an ongoing effort. Once started, it never ends. So the training provided in Step 6 is actually being provided early.

It has always been important to have well-trained, highly skilled employees. In the age of global competition, it is more important than ever. Poorly trained employees cannot deliver quality customer service. Companies must be proficient in assessing the need for ECS training and then for planning, coordinating, providing, and evaluating it.

TRAINING DEFINED

It is common to hear the terms "training" and "education" used interchangeably in discussions about enhancing a company's performance. Both concepts can contribute to continual improvement, but there are distinctions between them. In this book, "training" is defined as follows:

> ***Training** is an organized, systematic series of activities designed to continually enhance the performance of employees and, in turn, the company.*

Training has the characteristics of specificity and immediacy. Teaching and learning are geared toward the development of specific knowledge and skills that have direct application on the job.

Education typically describes teaching and learning in more formal settings where students must meet entrance requirements and teachers are required to have specific credentials, certification, and accreditation. Education is typically thought of as being more theoretical than training and having less specificity, immediacy, or direct application.

THE NEED FOR TRAINING

The need for training results from the demands on companies to be productive and competitive. This is why employers in the United States spend almost $50 billion per year on employee training and development at all levels. Several factors combine to intensify the need for training:

- the intensely competitive nature of the global marketplace
- rapid and continual change
- ever-increasing customer expectations
- changing demographics of the customer base

Unfortunately, and in spite of the huge sums spent on training every year, many employers still do not understand the important role of training. Less than 10 percent of U.S. employers use a flexible approach requiring better-trained workers as a way to improve performance. This approach is standard practice in Japan, Germany, Denmark, and Sweden. In addition, less than 30 percent of U.S. firms have special training programs for women, minorities, and immigrants, in spite of the fact that the majority of new employees comes from these groups. It is critical that employers understand the need for training.

For a company to succeed in a competitive global marketplace, every employee must be prepared to contribute ideas for improving performance. The best way to develop this capability is through constant training and retraining.

ASSESSING TRAINING NEEDS

What knowledge, skills, and attitudes do our employees need to outperform the employees of our competitors? What knowledge, skills, and attitudes do our employees currently have? The discrepancy in the answers to these questions points to training needs. A needs assessment may be necessary to determine this difference.

Companies can carry out needs assessments at the company, department, unit, team, or individual levels. Assessing training needs at the departmental level or below is not difficult. Supervisors work closely enough with their direct reports to see their capabilities firsthand. *Observation* is one method supervisors can use to assess training needs.

A more structured way of assessing training needs is to ask employees to evaluate (anonymously) their needs in terms of their job knowledge and skills. Employees know the tasks they must perform every day. They know which tasks they do well, which they do not do well, and which they cannot do at all. A *brainstorming* session focusing on training needs is another method companies can use.

The most structured approach companies can use is the *job-task analysis*, in which a job is analyzed thoroughly, and the knowledge, skills, and attitudes needed to perform it are recorded. Using this information, a survey instrument such as the one in Figure 6.1 is developed.

In responding to the survey, employees indicate which skills they have and which they need. The survey results can be used to identify (1) unitwide training needs and (2) individual training needs. Both types of needs are then converted into training objectives.

The following example demonstrates how a company might assess its training needs. Keltran, Inc. manufactures low-voltage power supplies for military and civilian aircraft. Production workers at Keltran install components on printed circuit boards, wrap wires on transformers, and build up wire harnesses for cables. The company receives many complaints from customers about quality problems with the power supplies. John Harris, production supervisor, is convinced that Keltran's production teams would benefit from specialized training aimed at improving performance. After mulling over the issue for a few days, Harris decides to assess the training needs of Keltran's three production teams: team A (printed circuit boards), team B (transformers), and team C (cables). Using what he learned from reading about needs assessment, Harris proceeds as follows.

1. Harris asks himself, "What are my most persistent production problems or bottlenecks, and what could be causing them?" He decides that incorrect components installed on printed circuit boards and improperly matched wires in cables are high on the list of frequently occurring problems. In his opinion, these problems are caused by a lack of blueprint-reading skills and an inability to adequately read technical manuals and specifications. He lists blueprint reading and general reading improvement as potential training needs.

2. Harris asks each team's lead employee to undertake the same process he has completed and to do so individually without discussing the issue among themselves. The lead employees list their perceptions of their team members' training needs and submit them to Harris.

3. Harris then calls a meeting of all production employees and explains that he is trying to assess their training needs and wants their input. After promising confidentiality, he distributes to each employee a form that has this statement typed across the top: "I could do my job better if training were provided for me in the following areas." All employees are asked to complete the statement and drop their forms into a box.

4. Harris analyzes the input he has received from the lead employees in each team and from the employees themselves. Adding his own, Harris compiles a master list of training needs. He gives highest priority to needs listed most often. They are as follows:

- general reading improvement

Job Knowledge and Skills Survey

The following is a list of tasks performed by people who work in the same job as yours or who work in related positions. In some cases, the task will be part of your job, and in other cases it will not. For each item on the list, circle the number that indicates how frequently, if at all, you perform this task yourself. If you do not understand the question, indicate this by circling #5.

	I perform this task				
	Never	Rarely, but only as a backup or in an emergency	Sometimes as a part of my regular job	Frequently as a part of my regular job	I don't understand this question
1. Use operating manuals, repair manuals, or other written instructions	1	2	3	4	5
2. Read blueprints	1	2	3	4	5
3. Interpret geometric dimensions and tolerances (*e.g.,* true positions, datums, flatness, circularity, or perpendicularity)	1	2	3	4	5
4. Read electrical schematics	1	2	3	4	5
5. Use charts or graphs to convert from one measure to another (*e.g.,* from inches to metrics) or to calculate speed and feed rates	1	2	3	4	5
6. Convert measures manually	1	2	3	4	5
7. Convert measures with a calculator	1	2	3	4	5
8. Use manually operated processing machines (*e.g.,* lathes, drill presses, sewing or fabric cutters)	1	2	3	4	5
9. Use computer numerically controlled processing equipment (*e.g.,* CNC milling machine, brake presses, or fabric cutters)	1	2	3	4	5
If yes, how many types of equipment?	1	2	3	4	5 or more

FIGURE 6.1 Sample training needs assessment instrument.

- interpretation of technical standards
- proper use of specifications
- blueprint reading

WRITING TRAINING OBJECTIVES

The next step in providing training for employees is to write training objectives—that is, objectives that specify in behavioral terms what skills are to be gained from the training program. This responsibility will fall in whole or in part to managers, supervisors, and other professionals. Some companies have training personnel who can assist. The key to success lies in being specific and stating objectives in behavioral terms. If a supervisor identifies a need for training in the area of math, she might write the following training objective: *Employees will learn mathematics.* This training objective both lacks specificity and is not stated in behavioral terms. Math is a broad concept. What does the training objective encompass? Arithmetic? Algebra? Geometry? Trigonometry? All of these? To increase the objective's specificity, the supervisor would need to break it down into several objectives.

To be stated in behavioral terms, training objectives must explain what the employee should be able to do after completing the training. Behavioral objectives contain action verbs. The sample training objectives in Figure 6.2 are stated in behavioral terms, are specific, and are measurable. The more clearly training objectives are written, the more effectively they can be used to plan training.

Sample Training Objectives

Upon completion of the training, employees should be able to:

1. Add, subtract, multiply, and divide whole numbers.
2. Add, subtract, multiply, and divide common fractions.
3. Add, subtract, multiply, and divide decimal fractions.
4. Convert everyday shop problems into algebraic expressions.
5. Solve right triangles.
6. Solve nonright triangles.

FIGURE 6.2 It is important to develop objectives for training.

PROVIDING TRAINING

Companies provide training in several different ways. These fall into one of the following broad categories: internal approaches, external approaches, and partnership approaches. Regardless of the approach used, companies always want to maximize their training resources. Strategies for doing this are as follows:

1. *Build in quality from the start.* Take the time to do it right from the outset.
2. *Be specific.* Do not develop courses that are all things to all people. Develop specific activities around specific objectives.
3. *Think flexible.* Do not automatically assume that the traditional classroom approach is best. Videotapes, interactive video, one-on-one peer training, or online training may be more effective.
4. *Be selective.* Before purchasing training services, conduct a thorough analysis of specific job-training objectives. Decide exactly what you want and make sure the provider you have in mind can handle it.
5. *Preview and customize.* Never buy a training product without previewing it. If you can reduce expense by customizing a generic product, do so.

Internal Approaches

Internal approaches involve providing training on site in company facilities. These approaches include one-on-one mentoring, computer-based training, media-based instruction, and group instruction. Mentoring involves placing a less-skilled employee under the instruction of a more-skilled employee. This approach often is used when a new employee is hired. It is also an effective way to prepare a replacement for a high-value employee who plans to leave or retire.

Computer-based training (CBT) has proven to be an effective internal approach, and it is now a widely used training method. It offers the advantages of being self-paced, individualized, and able to provide immediate and continual feedback to learners. It is a better application for developing general job-related knowledge than it is for developing company-specific job skills.

Formal group instruction, in which a number of people who share a common training need are taught together, is a widely used method. This approach could involve lectures, demonstrations, use of multimedia, hands-on learning, question and answer sessions, role playing, simulation, or any combination of these.

Media-based instruction has become a widely used internal approach. Private training companies and major publishing houses produce a list of turnkey media-based training programs. The simplest might consist of a set of videotapes; a more comprehensive package might include CDs and workbooks.

External Approaches

External approaches involve enrolling employees in offsite programs or activities provided by colleges, universities, professional organizations, and private training companies. The two most widely used approaches are: (1) enrolling employees in short-term training (a few hours to a few weeks) during work hours and (2) enrolling employees in short- or long-term training and paying all or part of the costs (*i.e.*, tuition, books, fees). External approaches range from seminars to college courses.

Partnership Approaches

In recent years, community colleges, universities, and technical schools have begun actively to pursue partnerships with employers, through which they provide customized training. Training partnerships combine some of the characteristics of the preceding two approaches.

Customized on-site training provided by universities, community colleges, and private companies or associations have become very common. Some colleges have built extensive networks of alliances with business, industry, and government employers.

Many universities, community colleges, and technical schools have continuing education or corporate training divisions that specialize in providing training for business and industry. It behooves companies to know the administrator responsible for training partnerships at the educational institutions in their communities.

Partnerships with institutions of higher education offer several advantages to companies. The representatives of these institutions are education and training professionals. They know how to transform training objectives into customized curricula, courses, and lessons; how to deliver instruction; how to design application activities that simulate real-world conditions; and how to develop a valid and reliable system of evaluation and use the results to chart progress and prescribe remedial activities when necessary. They also have access to a wide range of instructional support systems (*e.g.*, libraries, media centers, and instructional design centers).

In addition to professional know-how, institutions of higher education have resources that can markedly reduce the cost of training for a company. Tuition costs for continuing education activities are typically much less than those associated with traditional college-degree coursework. If these institutions do not have faculty members on staff who are qualified to provide instruction in a given area, they can usually hire a qualified temporary or part-time instructor.

Checklist
Advantages of Training Partnerships with Colleges, Universities, and Technical Schools

- Training professionals in colleges, universities, and technical schools know how to
  - transform training objectives into customized curricula, courses, and lessons
  - design learning activities that simulate real-world conditions
  - develop valid systems of evaluation
  - use the results of evaluation to chart programs and plan remedial activities
- Colleges, universities, and technical schools have access to instructional support systems, including libraries and media centers.
- Colleges, universities, and technical schools can offer credibility, formalization, credentialing, and standardization.

FIGURE 6.3 Training partnerships offer several advantages.

Other advantages offered by institutions of higher education in the training arena are credibility, formalization, standardization, and flexibility in training locations. Employers sometimes find their attempts at customized training are hampered because employees expect formal grades, transcripts, and certificates of completion. These proofs of education tend to formalize training in the minds of employees and can make it more real for them. Associating with a college, university, or technical school can formalize a company's training program and give it credibility.

Employers can also experience a lack of standardization when providing their own customized training. The same training provided in three different divisions might produce markedly different results. Professional educators can help standardize the curriculum and evaluation systems. They can also standardize instruction by providing train-the-trainer workshops for employees serving as in-house instructors. Figure 6.3 lists the advantages of training partnerships.

Regardless of the approach used in providing training, companies should remember this rule of thumb: *People learn best when their learning involves seeing, hearing, speaking, and doing.*

Educators hold that the following percentages apply regarding what learners are able to retain:

10 percent of what is read
20 percent of what is heard
30 percent of what is seen
50 percent of what is seen and heard
70 percent of what is seen and spoken
90 percent of what is said while doing what is being talked about

Clearly, for learning to be effective it must involve activity on the part of learners; must be interactive in nature; and must involve reading, hearing, seeing, talking, and doing.

EVALUATING TRAINING

Did the training satisfy the training objectives? This can be a difficult question to answer. Evaluating training requires that companies begin with a clear statement of purpose. This broad purpose should not be confused with training objectives. The objectives translate this purpose into more specific, measurable terms.

The purpose of training is to improve the individual performance of employees and the overall performance of the organization so that the organization becomes more competitive.

To understand whether training has improved performance, companies need to answer the following questions: (1) Was the training provided valid? (2) Did the employees learn what they were supposed to learn? (3) Has the learning made a difference? (Valid training is training that is consistent with the training objectives.)

Evaluating training for validity is a two-step process. The first step involves comparing the written documentation for the training (course outline, lesson plans, curriculum framework, and so on) with the training objectives. If the training is valid in design and content, the written documentation coordinates with the training objectives. The second step involves determining if the actual training provided is consistent with the documentation. Training that strays from the approved plan is not valid. Participant evaluations conducted immediately after completion of training can provide information on consistency and the quality of instruction. Figure 6.4 is an example of an evaluation instrument.

Determining if employees have learned is done by building an evaluation into the training. If the training is valid and employees have learned, the training should result in improved performance. Companies can determine if performance has improved using the same indicators that revealed the need for training in the first place.

Trainer:

Workshop Title:

Date: Month Day Year.

Instructions: On a scale from 5 to 1 (5 = highest rating to 1 = lowest rating), rate each item. Leave blank any item that does not apply.

Organization of Workshop **Comments:**

____ 1. Objectives (Clear = 5, Unclear = 1)

____ 2. Requirements (Challenging = 5, Unchallenging = 1)

____ 3. Assignments (Useful = 5, Not Useful = 1)

____ 4. Materials (Excellent = 5, Poor = 1)

____ 5. Testing Procedures (Effective = 5, Ineffective = 1)

____ 6. Grading Practice (Explained = 5, Unexplained = 1)

____ 7. Student Work Returned (Promptly = 5, Delayed = 1)

____ 8. Overall Organization (Outstanding = 5, Poor = 1)

Teaching Skills **Comments:**

____ 9. Class Meetings (Productive = 5, Nonproductive = 1)

____ 10. Lectures/Demonstrations (Effective = 5, Ineffective = 1)

____ 11. Discussions (Balanced = 5, Unbalanced = 1)

____ 12. Class Proceedings (To the Point = 5, Wandering = 1)

____ 13. Feedback (Beneficial = 5, Nonbeneficial = 1)

____ 14. Response to Students (Positive = 5, Negative = 1)

____ 15. Assistance (Always = 5, Never = 1)

____ 16. Overall Rating of Trainer's Teaching Skills (Outstanding = 5, Poor = 1)

Substantive Value of Workshop **Comments:**

____ 17. The course was (Intellectually Challenging = 5, Too Elementary = 1)

____ 18. The trainers command of the subject was
(Broad, Accurate = 5, Plainly Defective = 1)

____ 19. Overall substantive value of the course (Outstanding = 5, Poor = 1)

FIGURE 6.4 Form for evaluating instruction (to be completed by participants).

Can employees perform tasks they could not perform before the training? Has quality improved? Has customer satisfaction improved? Is the production rate up? Is the throughput time down? Companies can ask these types of questions to determine if training has improved performance. The following questions can be used for evaluating purchased training programs:

- Does the program have specific learning objectives?
- Does the program follow a logical sequence?
- Is the training relevant?
- Does the program offer opportunities for the trainees to apply the training?
- Does the program accommodate different levels of expertise?
- Is the philosophy of the program consistent with that of the organization?
- Is the trainer credible?
- Does the program provide follow-up activities to further develop job skills?

SUPERVISORS, MANAGERS, AND OTHER PROFESSIONALS AS TRAINERS

It is not unusual for in-house personnel to serve as instructors in a company-sponsored training program. In fact, this is becoming so common that it is important for a number of in-house personnel to have at least basic instructional skills. Supervisors need to understand the basic principles of learning and the four-step teaching approach.

Principles of Learning

The principles of learning summarize what is known and widely accepted about how people learn (see Figure 6.5). Trainers can facilitate learning more effectively if they understand the following basic principles:

- *People learn best when they are ready to learn.* You cannot make employees learn anything, but you can help them learn what they want to learn. Therefore, time spent motivating employees to learn is time well spent. Explain why they need to learn and how they will benefit personally from it.
- *People learn more easily when what they are learning can be related to something they already know.* Build today's learning on what was learned yesterday and tomorrow's learning on what was learned today. Begin each new learning activity with a brief review of the activity that preceded it.

Basic Principles of Learning

1. People learn best when they are ready to learn.
2. People learn more easily when what they are learning can be related to something they already know.
3. People learn best in a step-by-step manner.
4. People learn by doing.
5. The more often people use what they are learning, the better they remember and understand it.
6. Success in learning trends to stimulate additional learning.
7. People need immediate and continual feedback to know if they have learned.

FIGURE 6.5 Much of what is known about how adults learn is summarized by these principles.

■ *People learn best in a step-by-step manner.* An extension of the preceding principle, this means that learning should be organized into logically sequenced steps that proceed from the concrete to the abstract and from the broad to the specific.

■ *People learn by doing.* This is probably the most important principle. Inexperienced trainers tend to confuse talking (*i.e.*, lecturing or demonstrating) with teaching. Talking can be part of the teaching process, but it does little good unless it is followed with application activities that require the learner to do something. To illustrate this point, consider the example of teaching employees to roller-skate. You could present a thorough lecture on the principles of roller-skating and give a comprehensive demonstration of how to do it. Until the employees put on the skates and begin taking the first tentative steps, however, they have not begun to learn to skate. They learn by doing.

■ *The more often people use what they are learning, the better they remember and understand it.* How many things have you learned in your life that you can no longer remember? People forget what they do not use. This means that repetition and application should be built into the learning process.

■ *Success in learning tends to stimulate additional learning.* This is a restatement of a principle of management: "Success breeds success." Organize learning into segments that are short enough to allow learners to see progress, but not so short that they become bored.

Four-Step Teaching Approach

1. Preparation
2. Presentation
3. Application
4. Evaluation

FIGURE 6.6 Teaching is more than just lecturing.

- *People need immediate and continual feedback to know if they have learned.* Did you ever take a test and get the results back a week later? That was probably a week later than you wanted them. People want to know immediately how they are doing. Feedback can be as simple as a nod, a pat on the back, or a comment such as "Good job!" It can also be more formal, such as a progress report or a graded paper. Regardless of the form it takes, feedback should be provided on an immediate and continual basis.

Four-Step Teaching Approach

Teaching is a matter of helping people learn. One of the most effective approaches for facilitating learning is not new, innovative, or high tech in nature. Known as the four-step teaching approach, it is an effective approach in a corporate training setting (see Figure 6.6). The four steps are explained in the following paragraphs.

1. *Preparation* encompasses all the tasks necessary to prepare participants to learn, trainers to teach, and facilities to accommodate the process. Preparing participants means motivating them to learn. Trainers prepare themselves by planning lessons and preparing all the necessary instructional materials. Trainers prepare the facility by arranging the room for function and comfort, ensuring that all equipment works properly, and confirming that all tools and other training aids are in place.

2. *Presentation* is the act of presenting the material participants are to learn. It might involve giving a demonstration, presenting a lecture, conducting a question and answer session, helping participants work with a computer or interactive video-disk system, or assisting participants who are using self-paced materials. Regardless of the format, certain rules of thumb apply. The following strategies strengthen a presentation:

   - Begin dramatically (get learners' attention).
   - Be brief.
   - Be organized (use an outline and distribute copies).

- Use humor (laugh at yourself, not at the audience).
- Keep it simple.
- Take charge (be confident, but don't be arrogant).
- Be sincere.
- Be enthusiastic (let them see that you care).
- Tell stories to illustrate key points.

3. *Application* refers to providing learners with opportunities to use what they are learning. Application might consist of simulation activities in which learners role play, or hands-on activities in which learners use their new skills in a live format.

4. *Evaluation* is the step in which the trainer assesses the extent to which learning has taken place. In a training setting, evaluation does not need to be a complicated process. If the training objectives are written in measurable, observable terms, evaluation is simple. The employees demonstrate proficiency in fulfilling the objectives, and the trainer observes the results.

ECS TRAINING TOPICS

Any training that improves an aspect of an employee's performance is relevant and applicable in an ECS setting. There are certain topics that should be included in all ECS training programs, however. These topics represent skills that are important to all employees, regardless of position. They include the following:

- assertive listening
- dealing with difficult customers
- effective communication

The skills employees need to learn in these three fundamental areas are covered in Chapters 7 and 8. In addition to these fundamental training topics, there are several specific topics that should be offered to a more select group of employees. These topics include the following:

- conducting customer interviews
- facilitating focus groups
- designing questionnaires

CUSTOMER TRAINING

An old adage states, "The customer is always right." Although the intent of this message is a good one, in reality the customer is not always right. One of the most common reasons for consumer product failure is improper use by the customer. Typically, as many as one-third of all customer complaints result from improper use. This is why customer education is important.

Customer education, includes shaping customer expectations, providing user support, and marketing. To be satisfied with a product, customers need to know what to expect from it. This is important because in an ECS setting, quality is defined in terms of customer expectations. Customers are not likely to be satisfied if their expectations are inaccurate or unrealistic.

Customer expectations are shaped by the promotional literature used in marketing the product and by the user support materials provided with the product. For this reason, it is vital that promotional literature be accurate and that it not contain inflated claims. Accurate customer expectations can also be promoted by customer-service representatives. These employees should be adept at providing one-on-one training to customers, in person or by telephone. Toll-free numbers provide customers with easy access to customer-service trainers.

User support can be provided through user manuals, on-site technical assistance, or training provided at a company facility. Providing user support is an excellent way to train customers in the proper use of a product. To take full advantage of this opportunity, a company must provide readable user manuals, train its technical representatives to be customer trainers, and give customers immediate access to additional help through a user-support telephone number. User support can turn a new customer into a satisfied, knowledgeable, loyal customer.

Customer training can also help market a product. The philosophy that joins customer training and marketing can be stated as follows. "You would not buy a car if you did not know how to drive one." To derive the full marketing value of customer training, it's a good idea to involve marketing personnel in the development of the training.

Summary

1. Training is an organized, systematic series of activities designed to continually enhance the performance of employees and, in turn, the company. The need for training is driven by the following factors: the intensely competitive nature of the global marketplace, rapid and continual change, ever-increasing customer expectations, and the changing demographics of the customer base.

2. Commonly used methods for assessing the need for training include observation, brainstorming, and job-task analysis. Once the need for training has been determined, training objectives are written. Training objectives should be stated in specific behavioral terms that are measurable.

3. Training may be provided using internal approaches, external approaches, or partnership approaches. Regardless of the approach used, com-

panies should adopt the following strategies for maximizing their training resources: build in quality from the start, be specific, think flexible, be selective, and preview/customize.

4. Post-training evaluation helps to ensure that the training program satisfied the intended purpose. Companies need to know if the training provided was valid, if the employees learned what they were supposed to learn, and if it made a difference.

5. Supervisors, managers, and other professionals sometimes serve as trainers in their companies. They need to understand the basic principles of learning and the four-step approach to teaching (preparation, presentation, application, and evaluation).

6. ECS training topics that are important to all employees include assertive listening, dealing with difficult customers, and effective communication. Selected employees should also receive training on conducting customer interviews, facilitating focus groups, and designing questionnaires.

7. As many as one-third of all customer complaints are the result of improper use of the company's product. Consequently, it is important to provide training for customers in the proper use of the product. Customer training includes shaping customer expectations, providing user support, and marketing.

Key Phrases and Concepts

Application
Behavioral objectives
Computer-based training
Education
Evaluation
Feedback
Four-step teaching approach
Immediacy
Job-task analysis
Mentoring
Needs assessment
Preparation
Presentation
Principles of learning
Specificity
Training
Training objectives
Training partnership

Review Questions

1. Define the term "training."
2. What is the difference between education and training?
3. Why do organizations in the United States spend about $50 billion each year on education, training, and development?

4. Explain the impact of international competition relative to the training needs of organizations in the United States.
5. What is the most structured approach for assessing training needs?
6. Write a sample training objective that can be readily measured.
7. Explain briefly the following approaches to providing training: internal, external, and partnership.
8. What are the four principles of learning?
9. Explain the four-step teaching approach.

ECS APPLIED: DIVERSIFIED TECHNOLOGIES COMPANY PROVIDES ECS TRAINING

In the last installment of this case, David Stanley, DTC's CEO, informed the company's vice presidents for engineering (Meg Stanfield) and manufacturing (Tim Wang) of the resignation of their colleague and friend, Conley Parrish, the vice president for construction. Parrish resigned because he disagreed with the ECS philosophy. He plans to start his own construction company, a company that will compete with DTC. Stanley accepted Parrish's resignation with reluctance and regret. Meg Stanfield and Tim Wang updated Stanley on the various process-improvement projects underway in their departments.

"Meg and Tim, welcome Jake Arthur, our new vice president for construction," said David Stanley to kick off the meeting. "Jake, as you and I have already discussed, we are in Step 6 of a 10-step ECS implementation model. You have a lot of catching up to do, but Meg and Tim have agreed to help." "Thanks," responded the new vice president. "I've already met with Meg and Tim. They've been most helpful. The construction division will catch up right away. I am borrowing notes from Tim and Meg on everything that has happened in manufacturing and engineering so far. We can just replicate a lot of what has already been done." Turning to Meg Stanfield and Tim Wang, Arthur said, "I appreciate you folks plowing the ground for me. In fact, with your permission," said Arthur turning to Stanley, "I'd like to use customer interviews and a focus group to see if I can't hang on to some of the customers Conley plans to take from us. As much as they respect Conley, some of our customers have been complaining lately about poor customer service in the construction division." "Good idea," responded Stanley. "You have my permission. And one other thing, Jake. This might be a good time for me to tell you how we work in this little executive-management team of ours. Neither Tim nor Meg will admit it, but my vice presidents operate on the principle that it is easier to get forgiveness than permission. When you have what you think is a good idea, go with it. That's certainly what Tim and Meg do." Stanley smiled at his other two vice presidents who smiled back, but somewhat sheepishly.

"I'm glad you said that," acknowledged Arthur, "because we've already applied that theory with regard to training." "What do you mean?" asked Stanley. "Well," said Arthur, "When I came on board, Meg and Tim had already decided to combine their employees for the training sessions that are currently being offered on the topics of assertive listening, dealing with difficult customers, and effective communication. These topics apply equally to both engineering and manufacturing personnel, and I believe they also apply to our construction employees. So we have combined the employees from all three divisions for these training sessions." "Good job," said Stanley with enthusiasm. "I like your initiative."

Meg Stanfield then gave the group a brief report on the progress made in her division in identifying selected personnel to receive additional training on the topics of conducting customer interviews, facilitating focus groups, and designing questionnaires. Stanfield explained that after going over the list of key personnel several times, she had decided to include all supervisory personnel and above from her division in these training sessions. Wang and Arthur then explained that they agreed with Stanfield's decision and planned to do the same. "Good idea," agreed the CEO. "This will ensure that anyone who might be asked to interview a customer, facilitate a focus group, or help design a questionnaire will have had the necessary training. It will also make efficient use of the trainer's time. One more thing, as soon as one of our supervisors or managers gets the opportunity to conduct an interview or facilitate a focus group, let's have another join in as an observer. We'll do the same thing when we have a questionnaire to design. This will ensure that more of the folks who have the training get a chance to apply it." After some small talk on unrelated matters, Stanley wound up the meeting by saying, "I want the four of us to make sure that every employee gets the training on dealing with difficult and dissatisfied customers and starts using the training right away. And let's make sure that the four of us participate in one of the seminars and complete the training ourselves."

DISCUSSION CASES

The following cases provide examples of how the various concepts presented in this chapter might play out in actual companies. The cases are provided to prompt discussion, give the reader a feel for the types of problems confronted in the workplace, and reinforce the ECS concept in question.

CASE 6.1 We Don't Have Time for Training

The ABC Company is having trouble staying afloat. The competition is outperforming ABC on a daily basis. Its high turnover rate has caused the company to lose some of its best employees. New employees don't seem to have the knowledge, skills, or attitudes the company needs to improve its performance. ABC Company's CEO, Max Bender, is at his wit's end. He sees his company floundering, but he does not know what to do. A consultant brought in to help solve ABC's problems told Bender that the best strategy for getting his company on track could be summarized in three words: *Training, Training,* and *Training.* The consultant recommended training for new hires, retraining for experienced employees, and several seminars for top managers. Bender's response was emphatic. Just before dismissing the consultant he said, "We don't have time for training. While we are training, the competition will be running off with our customers. Besides, training is too expensive." Actually, ABC's most serious competitor had been training its employees in all aspects of ECS, which is why it was "running off" with ABC's customers. Within a year of dismissing the consultant, Max Bender initiated Chapter 11 bankruptcy proceedings for ABC Company.

Discussion Questions

1. It is widely believed in corporate America that the first budget to cut in hard times is the training budget. What is your opinion of this approach to cost cutting?
2. Defend or rebut the following statement: "During hard times we should increase our training."

CASE 6.2 Johnson Construction Company Assesses Training Needs

Bill Johnson had already completed a successful career as a civil engineer with the state department of transportation when he retired and used his "cash-out" bonus to finance the startup of his own business, Johnson Construction Company (JCC). JCC specializes in constructing roads, bridges, and parking lots. The competition is stiff in Johnson's state, but the CEO is confident that with his many years of experience he can make JCC the best-in-class. It was with this in mind that Johnson decided to pursue the concept of ECS. Johnson has already gone through Steps 1 through 5 of the ECS implementation model. There have been some rough spots, but essentially the implementation is going well. JCC has reached the point in the implementation process where employees at all levels need to undergo training.

Johnson has a general idea of the types of training that are needed, but he also understands that different managers and employees are at different places in terms of their ECS-related knowledge and skills. Consequently, Johnson has decided to conduct a needs assessment before launching the training program. To develop the needs-assessment survey instrument, Johnson has hired the same consultant who developed the company's customer questionnaire in Step 3 of the implementation process. The needs-assessment survey will contain questions divided into the following categories:

- listening skills
- conducting customer interviews
- facilitating focus groups
- questionnaire development and design
- dealing with difficult customers

After talking with the consultant, Johnson decides to develop more than one survey instrument. Not all employees will be expected to facilitate focus groups for example, but all do need to know how to listen well. Johnson will organize several working groups, made up of managers and employees, to work with the consultant in developing and pilot testing the survey instruments. Once the survey instruments have been developed, all employees will be asked to complete the instrument appropriate to their respective responsibilities. The tabulated results will inform Johnson as to specifically who needs what in terms of ECS training.

Discussion Questions

1. Has a company you have worked for conducted a training-needs assessment? Explain.
2. Think of a company where you have worked. What were the most critical training needs?

CASE 6.3 We Need Better Human Relations Skills

When Jack Poteet founded West Texas Contracting Company (WTCC), he brought together some of the best civil engineers and construction professionals available. Every member of his original team was a high performer in his or her area of specialization. Over the years, Poteet and the senior executive staff at WTCC maintained an *only-the-best* philosophy when filling critical positions. Unfortunately, the ability to work well with customers was not on the list of skills considered when hiring new employees. Jack Poteet now knows that oversight was a mistake. Over the past year, WTCC has lost three of its best customers as a result of what Poteet thinks of as "personality problems." Some of WTCC's best engineers and construction professionals just don't get along with people.

The lack of people skills among WTCC's key professional personnel is well known. In fact, it is part of the folklore that has grown up about WTCC among other engineering and construction firms in Texas. Once when making a speech to a civic club, Jack Poteet was asked about his company's reputation for cranky employees. He responded by saying, "My company is not in business to make friends. We make buildings and nobody does it better." This statement summed up the corporate culture at WTCC, a fact that after many years of success was now hurting the company.

Poteet knew he had to do something about building people skills in his company or WTCC was going to be knocked off its lofty perch in the marketplace. After some preparation on his own and consultation with an ECS consultant, Poteet called a meeting of his executive-level managers. He began the meeting by asking them a series of questions:

- How many of our employees have good telephone skills? How many of you do?
- How many of our employees know to handle an angry customer? How many of you do?
- How many of our employees have good conversation skills? How many of you do?

Poteet asked several more questions, all of which received the same response: "Not many." That's when he said, "It seems to me, we need better human relations skills." The CEO went on to explain that the three major customers WTCC had lost recently had all told him the same thing: "Jack, I just got tired of being treated like I don't matter. Your folks are good; no question about that. The problem is they think they're too good to be bothered with customers. I found a new company that can build my projects right, bring them in on time, and treat me and my people like we matter."

Poteet ended the meeting with the following message: "The rules of the game have changed. Just being good is no longer enough. We are going to have to learn human relations skills and use them when dealing with customers, or within a year or two we will all be standing in the unemployment line telling stories about the good old days."

Discussion Questions

1. In this case, Jack Poteet asked his executive-level managers three questions. Would you have added any questions? If so, what would they be?
2. Think of somewhere you have worked and apply the same questions to it. Are there any weaknesses that need to be corrected? Explain.

CASE 6.4 Floor Products Manufacturing and Pimco Construction Team Up to Train Customers

Floor Products Manufacturing (FPM) developed, patented, and now produces a line of flooring products that resemble wood, but never need to be waxed. Pimco Construction Products won the bid for exclusive rights to distribute FPM's "no wax" flooring products in a three-state region. In the beginning, the partnership was good for both companies. The "no wax" floor products were even more popular with customers than the most optimistic representatives of the company had anticipated. That was two years ago. Now things had changed, as if overnight, and not for the better.

Construction companies that had once stood in line to use the "no wax" floors were now dropping the products at an alarming rate. The problem was that within six months after installation, homeowners began to complain that their "no wax" floors need to be waxed.

Rather than panic, FPM and Pimco representatives, all of whom were sure the problem was not with the product, began visiting job sites where their floors were being installed. After just a few visits, it became apparent that faulty installation was the problem. Once installed, the floors must be sealed, and the sealant must be applied a certain way and in just the right amount. The contractors they observed had all properly installed the floors but had improperly applied the sealant.

Clearly, FPM and Pimco needed to train their customers—the local contracting companies who installed their "no wax" floor products. After a three-day conference at FPM's headquarters, the decision was made that Pimco would be staffed and equipped to provide installation training to contractors. In addition, FPM would install and staff a 1-800 number to receive customer complaints and give technical assistance.

Within just a few months, FPM and Pimco had begun to reclaim their lost customers. Within a year their "no wax" flooring products were once again market leaders. The difference between success and failure for an excellent line of flooring products was customer training.

Discussion Questions

1. Has any company you have dealt with provided customer training? If so, was it worthwhile?
2. Analyze the approach FPM and Pimco took in providing customer service. Would you have done anything differently? Why or why not?

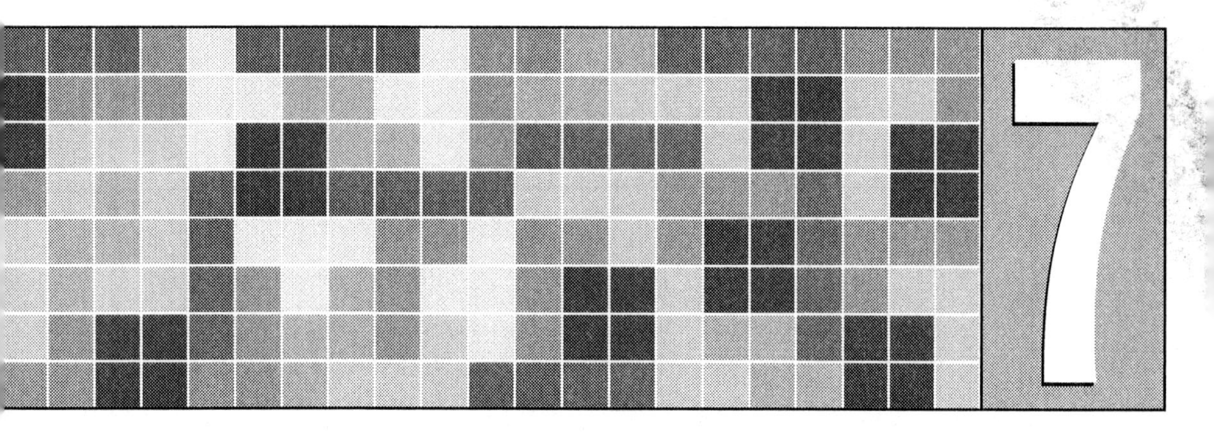

Turn Difficult and Dissatisfied Customers into Loyal, Repeat Customers

We tend to view customers who complain as problems, but in reality we should be thankful for them. Every time customers complain, they give us an opportunity to win their loyalty. Remember, they could just take their business elsewhere.

GOALS

- Understand the rationale for learning to deal with difficult customers.
- Discuss the reasons customers complain.

- Outline the seven-step model for handling customer complaints.
- Explain the types of treatment dissatisfied customers really want.
- State the absolute rule for handling angry customers.
- Turn dissatisfied customers into loyal, repeat customers.

Large retail establishments often have a separate department called "Customer Service," where unhappy customers can secure refunds or register complaints. The concept of the customer service department led one wag to ask the following pertinent question: *If your company has a customer service department, what do your other employees do?* The point of this question is that customer service is the responsibility of all employees. This is also a fundamental element of the ECS philosophy. In this chapter, we give companies the information they need to ensure that all their employees know how to deal with difficult and dissatisfied customers.

Engineering, manufacturing, and construction companies need all their employees to be positive agents when dealing with customers, especially when those customers are dissatisfied or difficult to deal with. When dealing with a complaining customer, it is easy for employees to become frustrated and say, "I don't have to listen to this." And it is tempting simply to write off customers who are difficult to deal with. The problem with this attitude is that it's not competitive. If you don't take the time to listen to your customers' complaints, your competitors will. Therefore, your employees need to know how to deal with difficult and dissatisfied customers and how to turn them into loyal, repeat customers.

RATIONALE FOR LEARNING TO DEAL WITH DIFFICULT CUSTOMERS

The best rationale for ensuring that all employees learn how to deal effectively with difficult customers is: *What you cannot or will not do, your competition will.* There are additional good reasons for learning to deal with difficult customers.

- *Dissatisfied customers talk.* Customers who take the time to complain are the exceptions. Most unhappy customers don't complain. Rather, they sign their next contract with one of your competitors. Worse yet, they talk about their dissatisfaction. It is common knowledge among customer-service professionals that an unhappy customer will tell at least 10 other people about his dissatisfaction.
- *Winning lost customers back is expensive.* Customers who migrate to the competition can be expensive to win back. A widely accepted ratio among customer-service professionals is that it costs five times as much to win back a customer as it does to retain the customer in the first place.

WHY CUSTOMERS COMPLAIN

When dealing with difficult customers, it helps to understand why they complain. One might think that the "why" would be obvious from the complaint, but this is not always the case. Some people are not good at communicating their grievances, especially when they are angry or upset. Others beat around the bush rather than come to the point. Still others are upset about one thing, but their frustration is compounded by yet other factors. It is not uncommon to have to read between the lines to determine what is really behind a customer's complaint. Fortunately, customer complaints are fairly predicable, at least in terms of the categories into which they can be generalized. Consequently, it is important to understand the categories of reasons that lead customers to complain.

Product or Service Lacks the Expected Quality

This is the worst customer service problem because it is so fundamental. Your company's products and services are the basis for its relationship with customers. When listening to customer complaints, employees should be particularly attentive to issues relating to product or service quality.

Difficulty Getting Through on the Telephone

Technology can be your company's best friend or worst enemy. After dealing with sophisticated computerized telephone-answering systems, many people throw up their hands and say, "How do I get through to a human being?" It is not uncommon for a customer to listen to all the various options and still not know which number to press for her problem. If your company uses a computerized answering system, ask someone who is completely unfamiliar with the company to call you. Then have that person critique the system for you. In addition, make sure the system includes an if-all-else-fails option that allows the caller to talk to a human being who can help. If your computerized telephone-answering system is frustrating your customers, it is not saving you money, no matter how good the budget figures look on paper.

Waiting

Most Americans are in a hurry, even when we don't need to be. We of the microwave generation are an impatient lot. We want, and can usually get, instant food, instant communication, instant cash, and instant information. As a result, we become frustrated when we are forced to wait. The longer we are forced to wait, the more frustrated we become. Being forced to wait can be the reason for a customer complaint, or it can be a factor that magnifies other frustrations. In either case, companies should be cognizant of the negative effects of forcing customers to wait, whether in person, on the telephone, or on line.

Rude Treatment

Few things turn a satisfied customer into a dissatisfied customer faster than being treated rudely. Every employee in your company should be trained to treat customers with respect and to maintain a positive attitude—even if the customer doesn't. We use the following analogy when training employees to deal effectively with difficult customers: *If you were hungry, you wouldn't be rude to someone who offered to feed you. Your company is hungry, and it is customers who feed it.*

Difficulty Understanding the Language

Engineering, manufacturing, and construction have their own languages. Customers may or may not be conversant with them. Consequently, your employees should be taught that whenever they communicate with customers—whether verbally, in writing, or electronically—they should speak the language of the customer, not of the business. The point is to communicate, not to show off professional vocabulary. If your customers have engineering, manufacturing, or construction backgrounds, speak to them in the language of the business. If not, use the language of lay people.

Listening Problems

Few people are good listeners, including those who think they are. Customers are typically very sensitive about whether they are being listened to—only rudeness annoys them more. Employers should assume that their employees are *not* good listeners and act accordingly, by providing training in the art and science of assertive listening for all employees (see Chapter 8).

HANDLING CUSTOMER COMPLAINTS

Two things are certain: (1) your company is going to receive complaints and (2) you cannot predict who in the company will actually receive the complaints. Consequently, all employees should learn how to handle customer complaints effectively. Figure 7.1 shows a seven-step model for handling customer complaints. The steps are explained in the following paragraphs.

Do Not Interrupt

Customers who complain are typically frustrated, angry, or both. It is good strategy to let the customer vent. Interrupting them will only make matters

Seven-Step Model for Handling Customer Complaints

1. Do not interrupt. Let the customer vent.
2. Listen attentively.
3. Paraphrase and repeat the complaint back to the customer (verification).
4. Offer a brief apology.
5. Discuss alternatives for solving the problem.
6. Secure customer confirmation.
7. Thank the customer.

FIGURE 7.1 It is important to handle customer complaints effectively.

worse. Problem solving is a rational, logical, participative process that cannot be done well when one of the participants is in an agitated frame of mind. Once a customer has been given the opportunity to vent without interruption, he usually settles down enough to participate effectively in the problem-solving process.

Listen Attentively

Listen attentively with your ears, your eyes, and your mind. When customers are upset, they may not express themselves well, and they tend to exaggerate. Listen to what is said—and also what is not said. Listen with your eyes (i.e., watch for nonverbal cues). It is important to note the differences between what is being said verbally and what is being said nonverbally. If there is incongruence, something is wrong. A good illustration of incongruence is a situation where a person says, "I am not lying to you" but is unable to look you in the eye. You should also listen with your mind. While listening and observing, apply your experience and intuition. Also, consider the typical reasons that customers complain, explained earlier in this chapter; this can help you sort out the facts from the emotional exaggerations.

Paraphrase and Repeat the Complaint

Once the customer has vented, paraphrase the complaint and repeat it back. This accomplishes two things, both of them positive. First, it shows the customer that you listened and that you are trying to discern what needs to be done. Second, it gives the customer an opportunity to make corrections if what you heard is not what was meant.

Offer a Brief Apology

One of the fastest, most effective ways to defuse anger and frustration in an upset individual is to apologize. Apologies work best when they are brief, sincere, and oriented toward solving the problem. Long apologies tend to sound patronizing and may actually anger rather than defuse the complaining customer. You can offer a simple apology even when your company is clearly not at fault. For example, you can say, "I am sorry about this trouble" or "I am sorry about the frustration this situation has caused you." In neither case have you said your company was wrong. Rather, you have apologized for the customer's frustration, which is real regardless of which party is in the wrong. No matter what you say by way of an apology, make sure that the last sentence of the apology is this: "Let's see how we can solve this problem." This sentence shifts the momentum from the complaint to the solution.

Discuss Alternatives

There is always more than one way to solve a problem. One option might be better for your company, and another might be better for the customer. The best solution is one that works for both. To find such a solution, discuss the alternatives that are available. When you discuss alternatives with the customer, you communicate an implicit message: "We are in this together, we are going to solve this problem, and you—the customer—have a voice in the solution." This message is critical because it is powerlessness that causes customers to feel angry and frustrated. Customers who are involved in selecting solutions from among various alternatives are more likely to support the eventual solution. Companies that determine the solution and then tell the customer what it is going to be are setting themselves up for another complaint. The solution may be rejected by the customer, or when it is implemented it may fail to meet the customer's expectations.

Get Customer Confirmation

After discussing solution alternatives with the customer, it is important to get confirmation that the chosen solution is acceptable. You can do this by making sure the customer understands all the ramifications of the solution. For example, assume a customer has lodged a complaint that 10 of the 100 units his company just purchased from yours are not operating properly. You discuss one alternative that involves repairing the units and another that involves replacing them. The customer likes the replacement option, but you need to let her know that to replace the units it is necessary to ship them in from an out-of-state plant. This will cause a seven-day delay in delivery. Therefore, the replacement option has both favorable and unfavorable aspects. The customer needs to know, understand, and accept both types of elements. The goal is to solve the problem, not exchange one problem for another.

Thank the Customer

One of the most important strategies for handling customer complaints is also one of the least used. It is simply to say, "Thank you." Customer complaints, even when delivered in an irritating manner, should not be viewed as irritants. Rather, as strange as it may sound, they should be viewed as gifts. A customer who complains is giving you the gift of the opportunity to keep his business. The appropriate response to such a gift is to thank the customer.

UNDERSTANDING WHAT DISSATISFIED CUSTOMERS REALLY WANT

To be effective at dealing with difficult and dissatisfied customers, one must know what they really want when they complain. It is also important to understand that dissatisfied customers want more than just a solution to their problem (see Figure 7.2). These factors are explained in the following paragraphs.

Acknowledgment/Validation

When a customer makes a complaint, she wants to be heard and taken seriously rather than being put off or ignored. Listening seriously to a customer's complaint is an acknowledgment that the customer is important and is being heard. Such an acknowledgment validates the customer's importance and lets her know that making a complaint is an appropriate and

Checklist: What Dissatisfied Customers Really Want

- Acknowledgment
- Validation
- Compensation
- Shared urgency
- Convenience
- Respect

FIGURE 7.2 Dissatisfied customers want more than just a solution to their problem.

appreciated option with your company. Failure to promptly and properly acknowledge and validate a customer only magnifies her dissatisfaction.

Compensation

When a customer goes to the trouble to complain, he wants your company to do more than just fix the problem. Fixing the problem just gets the customer back to where he started. Most customers expect some type of compensation for their time and trouble—they expect you to "go the extra mile." Consider the following example of going the extra mile. When the Volkswagen Group introduced the New Beetle, the car had an electrical defect that required a recall. In addition to fixing the problem, Volkswagen authorized dealers to spend $100 per customer as compensation for the customers' inconvenience. The money could be used to buy something for the customer, or it could simply be given to the customer in cash.

Shared Urgency

Some companies are guilty of taking their customers' complaints lightly. The attitude of personnel in such companies is, "This is your emergency, not mine." All employees in your company should understand that when a customer has a problem, the proper response is to immediately *adopt* the problem and share the customer's sense of urgency to get it solved. If the employee who receives the problem is not the right person to solve it, that employee should take responsibility for personally connecting the customer with the right person and making an introduction. The worst thing an employee can do is say, "That's not my department. You need to call . . . " Never tell a complaining customer what you cannot do or that his problem is not your problem. Rather, tell her what you can do. The customer has done her part by bringing the problem to your attention. From that point on, the responsibility for taking appropriate action rests with the company.

Convenience

Convenience is important to customers, and few things are more inconvenient than problems. Consequently, it's not enough just to fix a problem for a customer. He should also be reassured that the problem will not occur again.

Respect

Perhaps the greatest unstated need of a customer is the need to be treated with respect. Respect is a fundamental human need. There is a slogan that was once popular in customer-service circles that said, "The customer is always right." The problem with this slogan is that anyone who has worked with customers knows it's not true. Not only are customers not always right, they are frequently wrong. Because of this, the authors recommend a different slogan: "*Customers may not always be right, but they should always be treated right.*" Treating customers right means treating them with respect and giving them acknowledgment, validation, compensation, and shared urgency.

HANDLING ANGRY CUSTOMERS

In today's society, it is not uncommon for a customer to manifest his dissatisfaction in the form of anger. When people don't get what they want, it is increasingly the case that they respond with anger. Consequently, employees need to know how to deal with angry customers. Problem solving is a rational, logical process. Anger, on the other hand, is an irrational, illogical response.

There is one absolute rule for dealing with angry customers: *Remain calm and focus on the problem, not on the customer's anger.* Do not respond to anger with anger. This rule is easy to remember, but difficult to follow. Two strategies may help in this regard.

1. *Take a few deep breaths.* The normal physiological response to a stressful situation is increased heart rate. As the heart beats faster, one tends to become agitated. To counteract this response, take a few deep breaths. This will settle your heart rate and calm your nerves.
2. *Focus on the problem, not the anger.* One of the reasons another person's anger bothers us so much is that we tend to focus on the anger instead of the problem that is causing it. This is like a physician focusing on the infection around a splinter rather than on the splinter itself. Remove the splinter, clean the wound, and the infection will go away. It is the same with anger. Look beyond the anger and try to determine what is causing it.

Applying this rule requires practice and self-discipline, because the normal human responses to anger are either to return it in kind or to shrink away from it, neither of which is appropriate. Figure 7.3 is a list of behaviors to avoid when dealing with an angry customer. These responses are explained in the following paragraphs.

Do Not Agree with the Customer or Refer to Your Company as "They"

Even when the company is clearly wrong, employees need to play for the team they are on. Some people, when hearing a complaint, respond by passing the blame rather than taking responsibility for solving the problem. They characteristically just agree with the customer and refer to their employer as "they," saying such things as, "I know how you feel. They make this mistake all the time." Employees need to understand that when referring to

Behaviors to Avoid When Dealing with Angry Customers

DO NOT:

- Agree with the customer or refer to your company as "they."
- Become angry yourself and respond in kind.
- Turn your back on the customer and walk away.
- Hang up on the customer if on the telephone.
- Tell the customer that he is being rude or say, "I don't have to listen to this."

FIGURE 7.3 These behaviors will only serve to exacerbate a customer's anger.

their company, the proper term to use is "we." The proper action to take is to get the customer calmed down sufficiently to begin solving the problem.

Do Not Become Angry Yourself and Respond in Kind

The worst thing you can do when dealing with an angry customer is to respond in kind. This is like throwing gasoline on a fire, which is the opposite of what you want to happen. Remember that the first task is to calm the customer down so that you can work together in a logical, reasonable manner to solve the problem. When helping employees learn to deal with angry customers, ask them to write down the following slogan and memorize it: *When dealing with customers, if you lose your temper, you lose—period.*

Do Not Turn Your Back on the Customer and Walk Away

Some employees don't want to listen to an angry customer. Consequently, when a customer becomes angry, they simply turn around and walk away. It is important to distinguish between anger and abusive behavior. If a customer becomes abusive, it is appropriate—in fact recommended—for the employee to walk away and notify someone in authority. Judgment is required to determine whether you are in an abusive situation, but unless you feel threatened, stay with the customer and try to calm her down.

Do Not Hang Up on the Customer When on the Telephone

This issue is parallel to that of turning your back on a customer and walking away. The same advice applies in this case. Unless the customer crosses the line and becomes abusive, stay on the line and try to calm him down. Hanging up on a person is rude and just makes matters worse.

Do Not Say, "I Don't Have to Listen to This"

Some people respond to angry customers by pointing out that the customer is being rude. "You are being rude, and I don't have to listen to this." This response exacerbates the problem. The customer's response, whether spoken or unstated, is liable to be, "Maybe you don't have to listen to this, but I don't have to be your customer either, and I'll bet I can find another supplier who will listen to me."

To summarize, the most important task when dealing with an angry customer is to calm her down. You cannot begin to explore solutions until the customer is in a rational, logical frame of mind. The most effective way to calm an angry customer is to remain calm. Then, it is important to let the customer vent and to listen in a nonjudgmental manner. Once the customer has calmed down, the seven strategies in Figure 7.1 apply.

TURNING DISSATISFIED CUSTOMERS INTO LOYAL, REPEAT CUSTOMERS

The reason it is so important that all employees in engineering, manufacturing, and construction companies learn to deal with difficult and dissatisfied customers is that the ability to do so is one more factor that separates world-class companies from mediocre companies. Employees in world-class companies view dissatisfied customers as potential loyal, repeat customers. Figure 7.4 shows a five-step model for accomplishing this transition. The steps are explained in the following paragraphs.

Listen to the Complaint

This point has been stressed throughout this chapter. The three most important words to remember when dealing with a dissatisfied customer are: *listen*, *listen*, and *listen*. Acknowledge the customer and listen attentively to her complaint. Provide validation for the customer, share her urgency to get the problem solved, and treat her with respect.

Investigate the Complaint

Although it is important to treat the customer with respect, you do not want to simply accept his complaint at face value. Once you have all the necessary details, tell the customer you will look into the problem right

Five-Step Model for Turning Dissatisfied Customers into Loyal, Repeat Customers

1. Listen
2. Investigate
3. Act
4. Report
5. Follow up

FIGURE 7.4 A model for turning around unhappy customers.

away and get back to him. Tell him exactly when you will contact him, and then make sure you keep this promise. We recommend that companies teach their employees to *promise small and deliver big* when dealing with customer complaints. If you think you can get back to the customer by 2:30 P.M. today, tell him you'll call back by 3:30 P.M. and then call him at 2:30 P.M. The customer will think you have bettered your promise. Many people make the mistake of promising big but delivering small. This just makes the customer more dissatisfied. Investigate thoroughly, verifying all facts and claims. When attempting to solve problems, it is important to act on facts rather than on unverified claims or assumptions.

Act on What You Learn from the Investigation

When you are sure you have identified the source of the problem, solve it. If it turns out that the fault lies with the customer, take an objective look at whether your company could have done anything to make the transaction easier or more convenient for the customer. Regardless of what is found during the investigation, act on the facts to solve the problem, and do so as promptly as possible. The faster you solve a customer's problem, the greater your chance of winning her over. In this step, the company should go the extra mile to ensure customer satisfaction. Remember: Don't just solve the problem; compensate the customer for her inconvenience and trouble. If she ordered two, give her three or a discount on another product. If the products ordered got mixed up and she received the wrong batch, offer to pay the shipping costs on the next two orders. Dissatisfied customers have come to expect more than just a solution to the immediate problem. It is good business to meet this expectation.

Report Back to the Customer

One of the reasons people take their business elsewhere rather than investing the time and trouble to complain is that many have experienced a lack of follow-through when they have complained in the past. Not knowing the outcome of a complaint is actually worse than knowing that nothing happened. People like to know. Consequently, when dealing with dissatisfied customers, it is critical to keep them in the loop. If the problem has been solved, tell the customer. If it will take a couple days to solve the problem, tell him that, and give him an estimated completion date. If you have gotten the ball rolling on a fix for the problem but you are not sure when the solution will be finalized, tell him that. Then provide periodic updates.

Follow Up to Ensure that the Solution Had the Desired Effect

After you and the customer have discussed thoroughly the solution to be adopted, the eventual results may not turn out as planned. Consequently, it is important to follow up with the customer after the solution is in place to make sure that it worked as planned. If by chance the solution is not satisfactory, go back to Step 1 of the model in Figure 7.4 and try again.

Summary

1. The best rationale for ensuring that all employees learn how to deal effectively with difficult customers is this: What you cannot do or will not do, your competitors will. Additional reasons are: (1) dissatisfied customers talk and (2) winning lost customers back is expensive.

2. The most common reasons customers complain are lack of product or service quality, difficulty getting through on the telephone, waiting, rude treatment, difficulty understanding the language, or listening problems.

3. The following strategies can be helpful when handling customer complaints: Do not interrupt, listen attentively, paraphrase and repeat the complaint, offer a brief apology, discuss alternatives, get customer confirmation, and thank the customer.

4. Dissatisfied customers want more than just a solution to their problem. They also want acknowledgment, validation, compensation, shared urgency, convenience, and respect.

5. When dealing with an angry customer, the unbreakable rule is: Stay calm and focus on the problem rather than on the customer's anger. If you lose your temper and respond in kind, you lose—period. When dealing with

angry customers: Do not just agree with the customer and refer to your company as "they," do not become angry yourself and respond in kind, do not turn your back on the customer and walk away, do not hang up on the customer, and do not say, "I don't have to listen to this—you are being rude."

Key Phrases and Concepts

Acknowledgment/validation
Act on what you learn from the investigation
Compensation
Convenience
Difficulty getting through on the telephone
Difficulty understanding the language
Discuss alternatives
Dissatisfied customers talk
Do not become angry and respond in kind
Do not hang up on the customer
Do not interrupt
Do not just agree with the customer and refer to your company as "they"
Do not say, "I don't have to listen to this"
Do not turn your back on the customer and walk away
Follow up to ensure that the solution had the desired effect
Get customer confirmation
Investigate the complaint
Listen attentively
Listen to the complaint
Listening problems
Offer a brief apology
Paraphrase and repeat the complaint
Product or service lacks quality
Report back to the customer
Respect
Rude treatment
Shared urgency
Thank the customer
Waiting
Winning lost customers back is expensive

Review Questions

1. Explain the rationale for learning to deal with difficult customers.
2. What are the most common reasons customers complain?
3. List and explain each step in the seven-step model for handling customer complaints.
4. Defend or refute the following statement and fully explain your reasoning: "All a customer really wants when he has a complaint is for someone to solve his problem."

5. Defend or refute the following statement and fully explain your reasoning: "The customer is always right."
6. When dealing with an angry customer, how can you stay calm to avoid responding in kind?
7. What are the behaviors you should avoid when dealing with an angry customer.
8. Explain all the steps in the five-step model for turning dissatisfied customers into loyal, repeat customers.

ECS APPLIED: DIVERSIFIED TECHNOLOGIES COMPANY LEARNS TO DEAL WITH DISSATISFIED CUSTOMERS

In the last installment of this case, Meg Stanfield (vice president for engineering) and Tim Wang (vice president for manufacturing) updated DTC's CEO, David Stanley, on the ECS training projects underway in their divisions. Stanley was pleased with their reports, but still upset about the resignation of his long-time friend and colleague, Conley Parrish, vice president of construction at DTC. Parrish completely rejected the ECS philosophy and had refused to act on it.

"Jake, why don't you go first? How is the ECS implementation coming in your division? Are you about to get caught up?" asked David Stanley. "Thanks to Tim and Meg, we have almost caught up," answered Arthur. "By the time we meet again, the construction division will be on schedule with the implementation. I facilitated a customer focus group this week, which was an eye-opening experience. It went well, but I can tell you this—we had better not assume we know what our customers want." Wang and Stanfield laughed knowingly and agreed with Arthur. They both shared similar experiences they had with focus groups.

Arthur brought the group up to date on the various process-improvement projects underway in the construction division. Then he said, "What I most wanted to tell the group today is that the training we are conducting in dealing with dissatisfied customers is already paying off. About half the employees in my division have completed the training so far, and I got to observe one of them in action this morning." When David Stanley leaned forward in his chair expectantly, Arthur knew he had the group's attention. "Well, everyone likes a good story," thought Arthur, warming to the task and beginning to feel more comfortable with the group.

"Joe Acondi is one of our best and most experienced project superintendents," said Arthur. "Joe's been with the company since I was a teenager," acknowledged Stanley with a chuckle. "He's quite a character, but he's good."

"Yes, he is," offered Arthur. "But he's not one to suffer fools gladly. He is a tough, no-nonsense kind of guy who does not like to have his time wasted." "That's an understatement," said Meg Stanfield. "He was the superintendent when the company built my house. I made the mistake of stopping by to just chat one day. Vice president or not, he didn't have time for me. He asked me if I had anything to say that was germane to the project. But on the other hand, he brought the project in on time even though we had a hurricane right in the middle of the construction contract." "That's Joe," agreed Stanley with a laugh.

"Well I can guarantee you would not have recognized him this morning," offered Arthur. "He was on the Delano job site. That's that huge home we are building on Clearview Bay." "Nice home," interjected Wang. "It's a $2 million dollar project," said Arthur as an aside. "Anyway, Mr. Delano is one of those customers who thinks he knows more than he really does. He had been studying the plans and was convinced that our crew had laid out the foundation backwards." "You're kidding," laughed Stanley. "No, he was serious." "What did Joe Acondi do?" asked Stanfield, clearly expecting to hear about a blowup. "You won't believe it, but he was the very picture of patience," answered Arthur. "The only person who got angry was Delano. But the more flustered he got, the more patient Joe was with him. He visibly took a couple deep breaths to settle himself, and then he just let Delano vent. Once that was over, he apologized for the misunderstanding and showed Delano that the north arrow was oriented differently than it usually is. He handled the situation in a way that did not embarrass or even challenge Delano. Now Joe and Delano are big buddies. It was amazing. I was expecting bloodshed, but what I got was a lesson in how to deal with an angry customer. You would have been proud of Joe. I know I was." "That's one for the record books," said David Stanley, clearly pleased. "But it does show that our folks can learn to deal with dissatisfied customers. If Joe can do it, we all can."

DISCUSSION CASES

The following cases provide examples of how the various concepts presented in this chapter might play out in actual companies. The cases are provided to prompt discussion, give the reader a feel for the types of problems confronted in the workplace, and reinforce the ECS concept in question.

CASE 7.1 You Could Have Worked with This Customer If You Had Tried

Losing Can-Tech, Inc. as a customer was a major blow to Anderson Brothers Containers (ABC). Can-Tech, a leading manufacturer of aluminum cans and containers, had been one of ABC's best customers for almost 10 years. During this time, ABC produced the cardboard containers used to ship Can-Tech's products to customers throughout the world. The dispute that eventually led to the schism in this long-term business relationship was really just a clash of personalities.

When the long-time vice president for supplier relations at Can-Tech retired, Mark Donaldson, ABC's vice president for marketing, found himself working with a new contact. Mona McNamara, Can-Tech's new vice president for supplier relations, was a no-nonsense businessperson and a tough negotiator. Donaldson disliked her from the start. He had worked with her predecessor for 10 years and was accustomed to his laid-back, easy-going personality. McNamara's personality, in Donaldson's mind, was "too aggressive and obnoxious." The relationship between Donaldson and McNamara got off to a bad start, and it only went downhill with time. Rather than adapt himself to McNamara's interpersonal style, Donaldson rebelled against it. Tenuous from the outset, the relationship soon became downright hostile. Within six months of the change in vice presidents at Can-Tech, the 10-year relationship between ABC and Can-Tech was over.

When shortly thereafter he asked for Donaldson's letter of resignation, ABC's president said, "Mark, I've talked with Mona McNamara. I will grant you that she can be difficult, but she is not impossible. Had you tried, you could have found a way to work with her. I know because I did. We might get Can-Tech's business back some day. That remains to be seen. For now, because you didn't make the effort to learn how to deal with her, our company has lost its biggest and most important customer. That's unacceptable."

Discussion Questions

1. Have you ever had to work with a difficult customer? Explain.
2. Analyze how ABC's president handled this situation. Do you agree or disagree. Why?

CASE 7.2 Legitimate Complaint—Proper Response—Loyal Customer

"This could not have happened in a worse way or at a worse time," thought Steve Newton, CEO of Newton Construction Company. After two years of trying, Newton's company had finally won a contract with the Magna Group, a rapidly growing company that operates a popular chain of fast-food restaurants. Newton knew the contract was a test for his company. He also knew that if his company passed the test, there would be more contracts with the Magna Group—possibly many more.

Magna was adding restaurants throughout the world at a rate of 50 per year. If Newton Construction got on Magna's list of "certified" contractors, its business would triple within just two years. This is why the current problem was so frustrating. Newton Construction had finally won a "test contract" from Magna, only to face a delay in the groundbreaking ceremony because the secretary who handles the building-permit paperwork at the company was on vacation and had forgotten to file it with the county. With the groundbreaking ceremony coming up, Newton Construction did not have a building permit. The rule was simple: No building permit, no ceremony. This meant a delay of at least two weeks. Magna's executives are not happy, which is why Steve Newton decided to handle the problem personally.

Steve Newton met Magna's vice president for expansion projects, Juan Martinez, at the airport and escorted him to a limousine he had rented so they could begin discussions immediately. Newton wanted to begin the conversation by assuring Martinez that Newton Construction would set things right at no cost to Magna, but knowing it would be better to let Martinez vent, he bit his tongue and held back. He let the vice president do the talking, something Martinez was eager to do. Newton did not interrupt, but when Martinez appeared to have fully stated his complaint, the construction CEO paraphrased and repeated it back. "Let me make sure I understand everything you said," Newton offered. "Your company has already developed its marketing materials based on the original completion date for the project. Consequently, even though we are starting two weeks later, my company will still need to complete the project by the original date. In addition, all the dignitaries who planned to attend the groundbreaking ceremony will now have to be called and rescheduled, and they all have busy schedules. Finally, your company purchased nonrefundable airline tickets for the Magna executives who planned to attend the groundbreaking ceremony. This means you have lost approximately $5,000 in airfare expenses." "That's about the extent of it," replied Martinez, clearly exasperated.

Newton continued, "First, let me apologize on behalf of my company. I am sorry that this happened, and I am embarrassed about it. Now, let's see how we can solve this problem." Newton then proposed the following solution: (1) His company would guarantee the original completion date in spite of the late start and absorb any expenses necessary to meet the deadline, including overtime work; (2) Newton would personally call all the dignitaries who planned to attend the groundbreaking ceremony and reschedule the event for a date that would work for all of them and for Magna's executives. He would also apologize to the dignitaries and take full responsibility for the delay; and (3) Newton Construction would reimburse Magna for the funds it lost on the purchase of the original airline tickets, and it would purchase the new tickets for Magna's executives.

Martinez asked for clarification on one or two points, but then told Newton that the proposed solution was acceptable. Newton thanked Martinez for taking the time to bring him Magna's complaint in person and guaranteed that the project would proceed smoothly from that point on. Martinez was quiet for a few moments as if contemplating what he wanted to say. Finally he said, "Steve, I came here personally because I thought you would probably just offer excuses and try to wiggle out of your responsibilities. I am accustomed to that response when things go wrong. I don't mind admitting that I am pleasantly surprised and that I like the way you do business. If you make good on your promises and this project goes well, there will be others." "That's what I want," said Newton with a smile as he shook the Magna executive's hand.

Discussion Questions

1. Have you ever dealt with a representative of a company who made excuses rather than accept responsibility for poor quality?
2. Analyze how Newton handled this situation. Would you have done anything differently?

CASE 7.3 What's Wrong with Her? We fixed the problem.

Mark Brown could not understand why Vicki Robinson, vice president for supplier relations at Alpha Beta Corporation, was so upset. He took her telephone call, listened to her complaint, and got his people working on a fix right away. "What more does she want?" asked Brown, clearly exasperated. Mona Williamson, Brown's immediate supervisor, gave him a look that said, "Be patient, we'll talk this through." "It's not that you failed to solve the problem, Mark. You did a great job fixing the problem, and Alpha Beta's people appreciate what you did. So do I. They're not upset about what you did, but how you did it."

"What do you mean?" asked Brown. "Mark, I've learned over the years that when customers complain, they want more than just a solution to their problem," responded Williamson. "Solving the problem is important, in fact it's critical. If you don't do that, nothing else you do will matter. But you've got to do more than just solve the problem. You've also got to treat the customer who brings you a problem as if he is special." Williamson could tell from the expression on his face that Brown was either not getting it or not accepting it.

"Mark, let's run through the telephone call you had with Vicki Robinson. Begin with when you picked up the phone, and tell me everything that was said." "All right," mumbled Brown, unsure of where this conversation was headed. "I answered the telephone on the second ring. Vicki started telling me about her problem. She was going on and on, so I interrupted and asked her to cut to the chase. I told her I was really busy. When she finished telling me what she wanted, I told her I'd look into it and get back to her. That was it. Then I looked into the problem, found out what had happened, and got my people working on a fix right away."

"I think I see the problem, Mark. Vicki Robinson thinks you were short with her, in too big a hurry to get her off the telephone, and brusk in your attitude. Vicki thinks that since she represents one of our biggest customers, she should be treated better." "Come on, Mona," said Brown, now getting irritated. "Is this one of those male–female things?" "No, it's not," responded Williamson. "It's one of those customer–supplier things. I want you to think back to a couple of weeks ago when you had to put your car in the shop. Do you remember what happened?" "Of course I remember." Brown then recounted how he couldn't get anybody's attention. First, they let him stand around for more than 20 minutes before he finally had to stop someone and ask for help. Then they didn't know when they could get to his car and were unwilling to provide transportation to get Brown to work. Later in the day, when Brown called to check on his car, the telephone rang for a full five minutes before anyone answered it. Then, when a mechanic finally picked up the phone, he was rude and unhelpful. "Did the shop fix your car?" asked Williamson. "Yes, when they finally got around to it they did a good job." "But you are still unhappy with this particular shop?" queried Williamson. "I'll never take my car there again," said Brown with finality. "Mark, that's how Vicki Robinson feels about us right now."

Mona Williamson discussed several customer-service concepts with Brown, including acknowledgment, validation, compensation, shared urgency, and convenience. Brown was still a little defensive about his actions, but Williamson could tell the message was getting through.

Discussion Questions

1. Do you think customers have the right to expect more than just a solution to their problem? Why or why not?
2. Put yourself in Mark Brown's position. Would you have done anything differently? Explain.

CASE 7.4 Is Travel-Tech's Problem Solved?

Travel-Tech, Inc. (TTI) develops, designs, and manufactures convenience products for business travelers, especially those who travel by air. TTI's products range from eye shields that help people sleep on airplanes, to special clocks that help international travelers avoid jet lag, to battery-operated pillows that massage the neck. Electronic Products Company (EPC) has supplied the plastic and metal consoles for TTI's popular "Prevent-Jet-Lag" clocks for more than five years. TTI manufacturers the internal electronic components for the clock. A design change in the electronic components necessitated a corresponding change in the console. Unfortunately, an entire batch of the consoles had been manufactured according to the old design before EPC's engineers realized they had a problem.

When TTI's receiving and shipping department opened the box of outdated consoles from EPC, the quality control inspector noticed the problem right away. Because orders for the clock in question were already backlogged, the inspector lost no time in calling his contact at EPC. The EPC representative took full responsibility for the mix-up and promised to get right on the problem. Two days later, EPC's president and CEO, Cynthia Marcum, asked her vice president for manufacturing, Dave Weinstat, if the TTI problem had been solved. He didn't know, so he called his contact at TTI and asked. "No, we haven't heard a word from your people, and we need those consoles yesterday," was the response he got. When he called the manufacturing director for the project, Weinstat was told, "We are working on it. I can probably get it to TTI's receiving dock by noon tomorrow." Weinstat told the director to do whatever was necessary to ensure TTI received a double order of the correct consoles by noon the next day. Then he called EPC's business manager and told her to charge TTI for only one order. The second order was to be given to the customer free of charge.

Weinstat then called his contact at TTI and told him that a double shipment would arrive the next day, no later than 2:00 P.M. He also told him that there would be no charge for the second batch. This satisfied TTI's people. They could plan around a 2:00 P.M. shipment. The next day at precisely 2:00 P.M., Weinstat called his contact at TTI to inquire about the shipment of consoles. His contact was elated and said, "Dave, it got here at 11:45 A.M. Thanks for beating the deadline by a couple hours; that really helped us out. Now we are going to be able to fill some of our more important orders a couple of hours early. Oh, and thanks for the extra batch at no charge. That was a nice touch. I won't forget this next time we bid the console contract." Weinstat met later that week with EPC's president and CEO and recommended that she make mandatory for all departments training on how to deal with the type of problem they had with TTI. Cynthia Marcum agreed and said, "By the way Dave, I'd like you to teach the course."

Discussion Questions

1. Have you ever been promised something by a company that then failed to deliver on the promise? Explain.
2. How does the failure to deliver on a promise affect your image of a company?

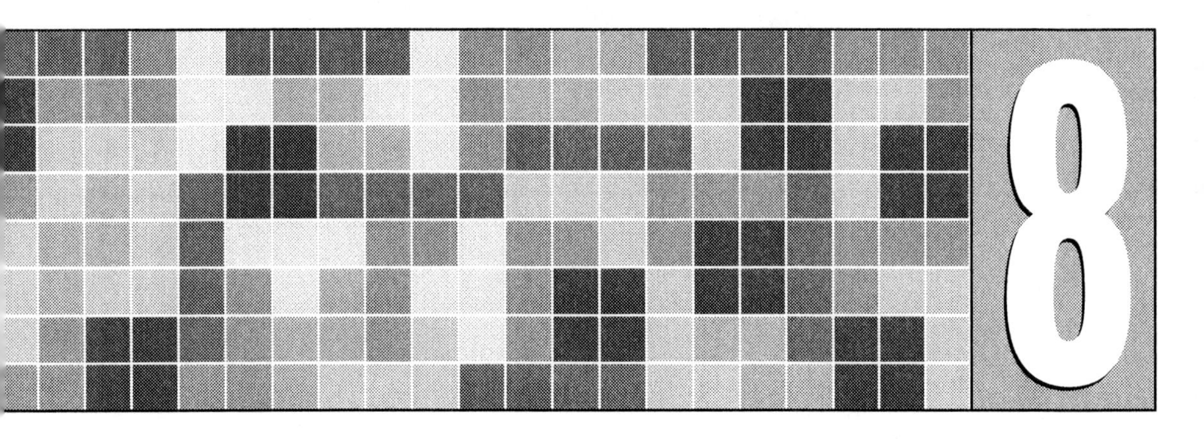

Communicate Effectively and Often with Customers

Effective communication is an important element in any successful relationship. This is as true with customer relationships as it is with personal relationships.

GOALS

- Define "communication."
- Understand communication as a process.
- Recognize inhibitors of communication.
- Establish a climate conducive to communication.
- Communicate by listening effectively.
- Communicate effectively verbally.
- Understand nonverbal communication factors.
- Communicate effectively in writing.

- Communicate effectively by telephone.
- Communicate effectively by email.
- Develop successful interpersonal skills.

Of all the skills needed in an ECS setting, communication skills are the most important. All the other strategies presented in this book depend either directly or indirectly upon effective communication. This chapter contains information about the communication and interpersonal skills engineering, manufacturing, and construction personnel need to communicate effectively with customers and colleagues.

DEFINING "COMMUNICATION"

People sometimes make the mistake of confusing "telling" with "communicating" and "hearing" with "listening." Then when a problem arises, they wonder why. You've probably heard this sentence, which illustrates that telling and hearing don't necessarily result in communication: "I know you believe you understand what you think I said, but I am not sure you realize that what you heard is not what I meant."

What you say is not necessarily what the other person hears, and what the other person hears is not necessarily what you intended to say. What is missing here is understanding. Communication may involve telling, but it is much more than that. Communication may involve hearing, but it is much more than that, too. "Communication" can be defined as follows:

> ***Communication** is the transfer of a message (information, idea, emotion, intent, feeling, or something else) that is both received and understood.*

A message may be sent by one person and received by another, but until the message is understood by both, no communication has occurred. This applies to all forms of communication—verbal, nonverbal, and written.

Effective Communication

When the message is received and understood, there is communication. However, communication by itself is not necessarily effective communication. *Effective* communication requires that the message be received, understood, and acted on in the desired manner.

For example, suppose a manager asks her team members to stay 15 minutes late every day for the next week, to ensure that an important order goes out on schedule. Each team member receives the message and verifies that he or she understands it. However, two team members decide not to comply. This is an example of ineffective communication. The two

Communication Levels in Engineering, Manufacturing, and Construction Companies

- Community level
- Company level
- Group level
- One-on-one level

FIGURE 8.1 Corporate communication can take place on various levels.

nonconforming employees understood the message, but decided against complying with it. The manager in this case failed to achieve acceptance of the message.

As stated above, effective communication involves receiving, understanding, and acting on a message in the desired manner. Reaching the standard of effective communication may require persuasion, motivation, monitoring, and leadership.

Communication Levels

Figure 8.1 shows the various levels at which communication typically takes place in a company. These levels are explained as follows:

- *One-on-one–level communication* involves one person communicating with one other person. This might involve face-to-face conversation, a telephone call, or even a simple gesture or facial expression.
- *Group-level communication* is communication within a group. The primary difference between one-on-one–level and group-level communication is that with the latter, all participants are involved in the process at once. A team meeting called to solve a customer problem or to set goals would be an opportunity for group-level communication, as would a meeting with several customers at once.
- *Company-level communication* is communication with customer companies. It occurs through such media as contracts, billing methods, and any other way a company projects its image to a specific customer company.
- *Community-level communication* occurs between a company and the community at large and consists of everything the company does to convey an image, market itself, and conduct its business.

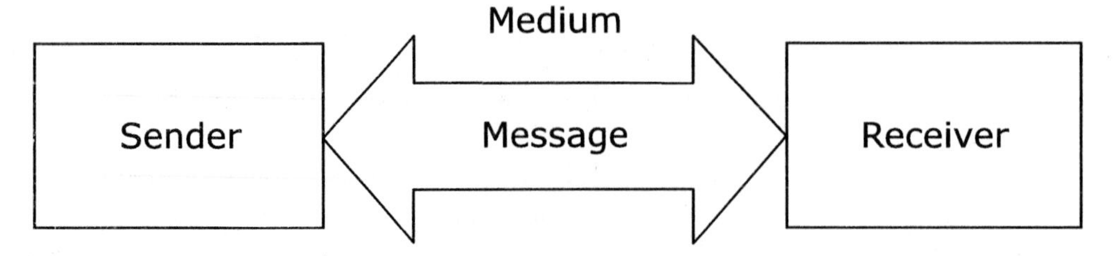

FIGURE 8.2 The communication process.

UNDERSTANDING COMMUNICATION AS A PROCESS

Communication is a process with several components–the message, the sender, the receiver, and the medium (see Figure 8.2). The *sender* is the originator or source of the message. The *receiver* is the person or group for whom the message is intended. The *message* is the information, idea, feeling, or intent that is to be conveyed, understood, accepted, and acted on. The *medium* is the vehicle used to convey the message.

There are four basic categories of media: verbal, nonverbal, written, and electronic. The *verbal* category includes face-to-face conversations, telephone conversations, speeches, public-address announcements, press conferences, and other means of conveying the spoken word. The *nonverbal* category includes gestures, facial expressions, physical appearance, and physical poses. The *written* category includes letters, emails, memoranda, billboards, bulletin boards, manuals, books, and any other method of conveying the written word. The *electronic* category includes the transmission of digital data as well as any other form of electronic transmission that can be converted into a message understood by humans (*e.g.*, the dot and dash impulses of a telegraph).

Technological developments, such as word processing, satellite communication, computer modems, cordless telephones, cellular telephones, answering machines, facsimile machines, pocket-sized dictation machines, and electronic mail, are having a major impact on our ability to convey information. No matter how advanced these communication-enhancing devices have become, there still exist as many inhibitors of effective communication as there ever were—maybe even more.

RECOGNIZING INHIBITORS OF COMMUNICATION

You should be familiar with the various factors that inhibit effective communication with customers. If properly handled, these inhibitors can be overcome or avoided. The most common inhibitors of effective communication are discussed below (see also Figure 8.3).

Inhibitors of Communication

- Differences in meaning
- Lack of trust
- Information overload
- Interference
- Condescending tone
- Poor listening skills
- Premature judgments
- Inaccurate assumptions
- Shoot-the-messenger syndrome

FIGURE 8.3 Many factors can make it difficult to have effective communication.

- *Differences in meaning.* People have different backgrounds, levels of education, and cultures. As a result, words, gestures, and facial expressions can have altogether different meanings for different people. This is why you should invest time in getting to know your customers.
- *Lack of trust.* If receivers do not trust senders, they may be overly sensitive and guarded. They might concentrate so hard on reading between the lines and looking for hidden agendas that they miss the message. This is why trust building with customers is so important.
- *Information overload.* Information overload is more of an inhibitor than it has ever been. Computers, modems, satellite communication, facsimile machines, email, and many other technological devices, which were developed to promote and enhance communication, can actually cause breakdowns in communication instead. Because of advances in communication technology and the rapid and continual proliferation of information, people often receive more information than they can deal with effectively. Be careful you don't overload customers with excessive information.
- *Interference.* Interference is any external distraction that inhibits effective communication. It might be something as simple as background noise or as complex as atmospheric interference with satellite communications. Regardless of the source, interference can distort or completely block out the message. You must therefore be attentive about the environment in which you plan to communicate.

- *Condescending tone.* People do not like to be talked down to, and they typically respond to tone of voice as much as or more than to the content of the message. It is a mistake to talk down to customers.
- *Poor listening skills.* Problems can result when the sender does not listen to the receiver and vice versa.
- *Premature judgments.* Premature judgments by either the sender or the receiver interfere with listening. When people make a judgment, they are prone to stop listening. One cannot make premature judgments and maintain an open mind. It is important to listen nonjudgmentally when communicating with customers.
- *Inaccurate assumptions.* Perceptions are influenced by assumptions. Inaccurate assumptions tend to shut down communication before it has a chance to get started. Don't assume; listen.
- *Shoot-the-messenger syndrome.* In the days when gladiators dueled in Rome's Coliseum, it was common practice to kill the bearer of bad news. A more civilized version of this practice is still very common, particularly in the workplace. If managers "shoot the messenger" when an employee tells the hard truth, they eventually hear only what employees think they want to hear. This dangerous situation can lead quickly to uninformed, ill-advised managers and poor customer service.

ESTABLISHING A CLIMATE CONDUCIVE TO COMMUNICATION

Corwin P. King describes a climate that is conducive to communication as one that "gives people the information they need to do their jobs well and also builds morale and encourages creativity."[1] He describes a bad communication climate as one that "creates doubt and confusion, demotivating people and leading to cynicism."[2] King goes on to describe how a bad communication climate can be guaranteed by the following behaviors:[3]

- communicating with peers, employees, and customers as little as possible, while at the same time being secretive and mysterious
- being vague and obscure: speaking in generalities
- communicating with only a select few individuals
- limiting employee and customer access
- communicating only when it is personally advantageous to do so
- ignoring employees' and customers' good ideas

If these are the steps to creating a bad communication climate, the obverse of each step should create a climate that is conducive to communication. Therefore, you should take special care to communicate specific, detailed information with as many employees and customers as possible, as frequently as possible, and at the same time solicit their ideas for improvement.

COMMUNICATING BY LISTENING

Listening is one of the most important communication skills. It is also the one people are least likely to have. Are you a good listener? To find out, complete the listening skills assessment in Figure 8.4.

(If you are a good listener, your answers to the questions in Figure 8.4 will be: 1, no; 2, no; 3, no; 4, yes; 5, no; 6, no; 7, yes; 8, no; 9, no; and 10, no.)

Good listeners typically listen more than they talk. Interrupting people before they complete a statement is a sign of impatience and lack of interest in what is being said. This can have a doubly negative effect on communication. First, interrupting a speaker lessens the listener's chances of properly perceiving what is being said. Second, it sends the message "I don't have time to listen to you."

People who tune out and think ahead to their response have more interest in their own message than the speaker's. Accurate perception is difficult enough when the listener is tuned in; it is impossible when he is tuned out. A tuned-in listener should be able to digest a speaker's message, paraphrase, and repeat it back. Paraphrasing does not mean parroting back the speaker's exact words. It means summarizing the message in one's own words, to let the speaker know she has been heard and understood.

Stating an opinion before a speaker has finished her message or continuing other tasks while someone is speaking both send the same message: "I don't want to hear it." People who send this message get what they ask for—the speaker stops trying to communicate. The problem with this is that in a competitive environment, the customer's message needs to be heard.

Even the best listeners sometimes need to ask questions for clarification. Questioning not only improves accuracy of perception, it also shows the speaker that the listener is tuned in and wants to understand. There are two ways to handle questions. The first is to wait until the speaker pauses or begins to move on to another point, and then to raise the question; doing so at such a juncture will not cause the speaker to lose his train of thought. If it is critical that a point be clarified immediately, however, it is acceptable to stop the speaker by raising a hand in a gesture meaning, "Hold on a moment." If you stop a speaker in this way, make a mental or written note of where he left off, in case a reminder is needed to get the conversation started again.

Daydreaming during meetings and sneaking glances at your watch both say: "I've got something better to be doing with my time." Time pressures and conflicting demands for your time are realities, but few things are as

ECS Listening Self-Assessment

Question	Response	
	Yes	No
1. When with a customer, do you talk more than you listen?		
2. When talking with a customer, do you frequently interrupt?		
3. In conversations with customers, do you tune out and think ahead to your response?		
4. In a typical conversation with a customer, can you paraphrase what the speaker has said and repeat it?		
5. When talking to customers, do you state your opinion before they have made their case?		
6. Do you continue doing other tasks when a customer is talking with you?		
7. Do you ask for clarification when you don't understand what a customer has said?		
8. Do you frequently tune out and daydream during meetings with customers?		
9. Do you fidget and sneak glances at your watch during conversations with customers?		
10. Do you find yourself finishing statements for customers who don't move the conversation along fast enough?		

FIGURE 8.4 Learn how well you listen.

Inhibitors of Effective Listening

- Lack of concentration
- Preconceived ideas
- Thinking ahead
- Interruptions
- Tuning out
- Interference

FIGURE 8.5 These factors and behaviors can cause ineffective listening.

important as listening to customers. People who find themselves tuning out in meetings and glancing at their watches during conversations should give thought to how they are organizing and managing their time.

Finishing sentences for people you feel don't move the conversation along fast enough sends the message: "I am in too big a hurry to listen to you." Of course, this might legitimately be the case. If so, rather than finishing sentences for a customer, you might get better results by saying, "I cannot give you the attention you deserve right now. Let's compare schedules and find a time that works for both of us."

What Is Listening?

Hearing is a natural process, but listening is not. A person with highly sensitive hearing capabilities can be a poor listener. Conversely, a person with impaired hearing can be an excellent listener. *Hearing* is the physiological decoding of sound waves, whereas *listening* involves perception. "Listening" can be defined in numerous ways. We use the following definition:

> ***Good listening*** *means receiving a message, correctly decoding it, and accurately perceiving what it means.*

Inhibitors of Effective Listening

Effective listening occurs when the receiver accurately perceives a message's meaning. Unfortunately, several inhibitors can prevent this from happening. These inhibitors include the following (see Figure 8.5):

- lack of concentration
- interruptions

- preconceived ideas
- thinking ahead
- interference

To perceive a message accurately, listeners must concentrate on what is said, how it is said, and in what tone it is said. Effective listening also involves properly reading nonverbal cues (discussed in the next section).

Interruptions not only inhibit effective listening, they can also frustrate and confuse the speaker. If clarification is needed during a conversation, it is best to make a mental note and wait for the speaker to reach an interim stopping point. Mental notes are always preferable to written notes. The act of writing notes may distract the speaker—or the listener. If you find it necessary to make written notes, keep them short and to a minimum.

Concentration requires that the listener eliminate as many distractions as possible and mentally shut out the rest. People who hold preconceived notions most often do not listen effectively. Preconceived ideas can lead to premature judgments that turn out to be wrong. Even the most experienced communicators benefit from waiting patiently and listening.

People who jump ahead to where they think the conversation is going often arrive there only to find they are alone. Thinking ahead is typically a response to being hurried; it is actually the case, however, that it takes less time to hear a customer out than it does to start over after jumping ahead in the wrong direction.

Some people become skilled at using body language that makes it appear they are listening when, in reality, their mind is focused elsewhere. Avoid the temptation to engage in such ploys. Skilled speakers may ask you to repeat or paraphrase what they have just said.

Interference is anything that distracts the listener, impeding either hearing or perception or both. Background noises, a telephone ringing, and people walking through the office are all examples of interference. Such distractions should be foreseen and prevented before beginning a conversation. If they cannot be, the conversation should be moved to another location. Figure 8.6 provides a checklist you can use to improve your listening skills.

Listening Assertively

One way to improve your ability to properly perceive messages is through assertive listening.

> ***Assertive listening*** *means listening with your ears, eyes, and mind in an attempt to fully understand what is meant by the message.*

Assertive listening is the highest level of listening, the first three levels being hearing but ignoring, pretending to listen, and selective listening.

Listening Improvement Checklist

- ☑ Remove all distractions.
- ☑ Put the speaker at ease.
- ☑ Look directly at the speaker.
- ☑ Concentrate on what is being said.
- ☑ Watch for nonverbal cues.
- ☑ Make note of the speaker's tone.
- ☑ Be patient and wait.
- ☑ Ask clarifying questions.
- ☑ Paraphrase and repeat.
- ☑ No matter what is said, control your emotions.

FIGURE 8.6 It is important to continually improve your listening skills.

Ignoring the message and the messenger can be done overtly or covertly. Overt signs of ignoring include interrupting, fidgeting, and stealing glances at a watch. Covertly ignoring the message means hearing the messenger but completely disregarding what she says. *Pretending* means acting as if you are listening when you are actually tuning out. Pretenders use nonverbal cues, such as eye contact, nods, and facial expressions, to give the impression they are listening. *Selective listening* means tuning in to only parts of the message. It's like viewing a painting that is partially covered: The best part may be what you don't see.

Assertive listening is attentive listening; it involves focusing on the speaker's words while giving equal attention to nonverbal cues, feelings, emotions, intensity, and so on. To perceive messages fully, it is often necessary to do more than just listen to the words.

Improving Listening Skills

Most people have room for improvement in their listening skills. Fortunately, these skills *can* be improved, particularly when there is an awareness of the need to improve. C. Glenn Pearce recommends the following strategies:[4]

- *Upgrade your desire to listen.* Customer service requires that you listen more and talk less. Many people need to make a concerted and conscious effort to suppress their natural desire to talk—the verbal equivalent of sitting on one's hands. A good strategy is to make a conscious effort to learn as much as possible from every conversation. This forces the issue of listening instead of talking.

■ *Ask the right questions.* Two people can hear the same words but receive different messages. Consequently, it is important to ask questions that clarify the message. Three types of questions can be helpful in this regard. The first is used to move the speaker on to his next point ("I understood your first concern, is there a second?"). The second is used to gain an intermediate summary of the conversation before moving on to a new point ("Can you summarize this concern before we move to the next one?"). The third type of question is used to obtain a summary of the entire conversation ("Before leaving, can you summarize your major concerns for me?").

■ *Judge what is really being said.* This involves listening to more than just words. It means observing nonverbal cues, rate of speech, tone of voice, intensity of the speaker, enthusiasm or a lack of it, and context clues, to get beyond what is said to why it is being said.

■ *Eliminate listening errors.* Listening errors include failing to concentrate, tuning out, giving in to distractions, and interrupting. To eliminate these errors, keep tabs on how frequently you make them. For a given week, jot down the number and type of listening errors you made during each conversation you had. This will boost your awareness and help you correct the listening errors you make most frequently.

UNDERSTANDING NONVERBAL COMMUNICATION FACTORS

Communications consultant Roger Ailes explains the importance of nonverbal communication as follows: "You broadcast verbal and nonverbal signals that determine how others see you. . . . And whether people realize it or not, they respond immediately to your facial expressions, gestures, stance, and energy, and they instinctively size up your motives and attitudes."[5] Figure 8.7 lists the components of nonverbal communication.

Body Factors

Posture, body poses, facial expressions, gestures, and dress can convey a message. Even such extras as makeup or the lack of it, well-groomed or unkempt hair, and shined or scruffy shoes can convey a message. You should be attentive to these body factors and how they add to or distract from the verbal message.

One of the keys to understanding nonverbal cues lies in the concept of congruence. Are the spoken message and the nonverbal message consistent? They should be. Incongruence occurs when words say one thing, but non-

Components of Nonverbal Communication

Body Factors
- ☑ Posture
- ☑ Dress
- ☑ Gestures
- ☑ Facial expressions
- ☑ Body poses

Voice Factors
- ☑ Volume
- ☑ Pitch
- ☑ Tone
- ☑ Rate of speech

Proximity Factors
- ☑ Relative positions
- ☑ Physical arrangements
- ☑ Color of the room or the environment
- ☑ Fixtures

FIGURE 8.7 Nonverbal communication includes more than just body language.

verbal cues say another. When the verbal and nonverbal aspects of the message are incongruent, dig a little deeper. An effective way to deal with incongruence is to confront it gently, with a simple statement such as: "Mary, your words say one thing, but your nonverbal cues say something else."

Voice Factors

Voice factors are an important part of nonverbal communication. In addition to listening to the words in a message, you should listen for such factors as volume, tone, pitch of voice, and rate of speech. These factors can indicate anger, fear, impatience, unsureness, interest, acceptance, confidence, and a variety of other feelings. As with body factors, it is important to look for congruence. It is also advisable to look for groups of nonverbal cues.

People can be mislead if they attach too much meaning to isolated nonverbal cues. A single cue taken out of context may have little meaning, but as part of a group of cues, it can take on significant meaning. For example, if you look through an office window and see a man leaning over a desk pounding his fist, it would be tempting to interpret this as a gesture of anger. But what kind of expression does he have on his face? Is it congruent with desk-pounding anger? Or could he simply be trying to knock loose a stuck desk drawer? If he is pounding on the desk with a frown on his face and yelling in an agitated tone, your assumption of anger might be correct. He might be angry just because his desk drawer is stuck, but he is still angry.

Proximity Factors

Proximity involves a range of factors such as where you position yourself when talking with an employee, how your office is organized, the color of your walls, and the type of fixtures and decorations. Sitting next to a customer conveys a different message than sitting across a desk from her. A person who goes to the trouble of making his office a comfortable place to visit sends a message that invites communication. A person who maintains a stark, impersonal office sends a less welcoming message.

To send the nonverbal message that customers are welcome, try using the following strategies:

- Have comfortable chairs available for visitors.
- Arrange chairs so you can sit beside visitors rather than behind your desk.
- Choose soft, soothing colors rather than harsh, stark, or overly bright colors.
- If possible, have refreshments such as water, coffee, and soda available for visitors.

COMMUNICATING VERBALLY

When dealing with customers, verbal communication ranks close in importance to listening. You can improve your verbal communication skills by emphasizing the attributes listed in Figure 8.8 and explained in the following paragraphs.

- *Show interest.* When speaking with customers, show an interest in the topic of conversation. Show that you are sincerely interested in communicating your message to them. Also, demonstrate an interest in the receivers of the message themselves. Look them in the eye, and when in a group, spread your eye contact evenly among all receivers.

Strategies for Improving Verbal Communication Skills

- Show interest.
- Be friendly.
- Be flexible.
- Be tactful.
- Be courteous.

FIGURE 8.8 Work on improving your verbal communication skills every day.

- *Be friendly.* A positive, friendly attitude enhances verbal communication. A caustic, superior, condescending, argumentative, or disinterested attitude will shut off communication. Be patient, be friendly, and smile.
- *Be flexible.* Flexibility can enhance verbal communication. If you as a manager call your employees together to discuss a customer issue, but find they are uniformly focused on a problem that is disrupting their work schedule, be flexible enough to put your message aside for the moment and deal with their problem. Until the employees work through what is on their minds, they will not be good listeners.
- *Be tactful.* Tact is an important ingredient in verbal communication, particularly when one is delivering a sensitive or potentially controversial message. Tact has been called the ability to hammer in the nail without breaking the board. The key to tactful verbal communication lies in thinking before speaking.
- *Be courteous.* Being courteous means showing appropriate concern for receivers' needs. Calling a meeting 10 minutes before quitting time is discourteous and will inhibit communication. Courtesy would also prevent one from monopolizing a discussion. When communicating verbally, give receivers ample opportunities to ask questions, seek clarification, and state their points of view.

Asking Questions Effectively

Communication skills can also be strengthened by learning to be a skilled questioner. Knowing how and when to ask questions is an important verbal communication skill. It helps you get at what customers really think and feel. Some general rules of questioning used by professional counselors to draw out their clients' feelings and thoughts apply here. You can apply these same rules in your verbal communication with customers (see Figure 8.9).

- *Drop your defenses.* Human interaction is emotional interaction. There is no such thing as fully objective discourse between people. All people have their public and private faces; rarely does what is said completely match what is felt. People learn early in life to put up defenses. To communicate effectively, it is necessary to break through the defenses. A strategy counselors use is to drop their own defenses first. When customers see you open up, they are more likely to follow suit and respond more openly to your questions.
- *State your purpose.* The question people often ask themselves when they are asked a question is "Why is she asking that? What does she really want?" You will learn more from your questions if you state your purpose at the outset. This allows the receiver to focus on the question itself rather than worrying about your motivation for asking it.
- *Acknowledge emotions.* Avoid what counselors call the "elephant-in-the-living room" syndrome. Human emotions can be difficult to deal with, and as a result, some people respond by ignoring them. This is like walking around an elephant in the living room and pretending you don't see it. Ignoring the emotions of people you question may cause them to close up. If a person shows anger, you might respond by saying, "I can see I've made you angry" or "You seem to feel strongly about this." Such nonjudgmental acknowledgments often draw people out.
- *Use open-ended questions and phrase questions carefully.* To learn the most from your questions, make them open-ended. This allows the person being questioned to do most of the talking, and in turn, you to do most of the listening. Counselors feel they learn more when listening than when talking. Closed-ended questions force restricted or limited responses. For example, the question "Can we meet our deadline?" will probably illicit a yes or no

Questioning Strategies

- Drop your defenses.
- State your purpose.
- Acknowledge emotions.
- Use open-ended questions.
- Phrase questions carefully.

FIGURE 8.9 Strategies for effective questioning enhance verbal communication.

response. However, the question "What do you think about this deadline?" gives the responder room to offer opinions and other potentially useful information.

Practice using these questioning techniques at home, at work, and even in social settings. It takes practice to internalize them to the point that they become natural. With practice, however, you can become a skilled questioner and, as a result, a more effective communicator.

COMMUNICATING IN WRITING

The ability to communicate effectively in writing is an important customer-service skill. Like any other skill, the types of writing required can be mastered with the appropriate mix of coaching, practice, and genuine effort. This section provides the coaching. You have to provide the practice and effort yourself.

Helpful Rules

Several rules of thumb can enhance the effectiveness of your written communication. They are explained in the following sections and listed in Figure 8.10.

- *Plan before you write.* One barrier to effective writing is starting to write before deciding what you want to say. This is like getting in a car and starting to drive before deciding where you want to go. The route is sure to be confusing, as will the message if you write without planning first.

Planning a memorandum or letter is a simple process. It is a matter of deciding who you are writing to, why, and what you want to say. Figure 8.11 is a planning sheet that will help you plan before you write. Taking the

Strategies for Better Writing

- Plan before you write.
- Be brief.
- Be direct.
- Be accurate.
- Practice self-editing.

FIGURE 8.10 Rules of thumb for enhancing written communication.

Planning Form for Letters and Memoranda

- I am writing to ______________________________

____

____

____

- My purpose in writing is to ______________________________

____

____

____

- I want to make the following points: ______________________________

A. ______________________________

B. ______________________________

C. ______________________________

D. ______________________________

- I want the recipient to do the following: ______________________________

____

____

____

FIGURE 8.11 Plan first, write second.

time to complete such a planning sheet helps ensure that you communicate your intended message. When you have used planning sheets for a time, you will be able to perform this step mentally without having to actually fill them out.

- *Be brief.* One of the negative aspects of the modern technological environment is the potential for information overload. This relates to another negative aspect—the tendency toward shortened attention spans. Modern computer and telecommunication technology have conditioned customers to expect instant information with little or no effort on their part. Written documents can run counter to this expectation of instant information with no effort. Reading takes time and effort. Keep this in mind when writing and be brief. In as few words as possible, explain your purpose, state your points, and tell recipients what you want them to do.

- *Be direct.* Directness is an extension of brevity. It means getting to the point without beating around the bush. This is especially important when the message is one the reader will not particularly like. No purpose is served by obscuring the message. Come right to the point and state it completely and accurately.

- *Be accurate.* Accuracy is important in written communication. Be exact. Avoid vague phrases such as "some time ago," "approximately," and "as soon as possible." Take the time to identify specific dates, numbers, quantities, and so on. Then double-check to make sure they are accurate before making your written communication public.

- *Practice self-editing.* The one-draft writer is rare. But it is very common that people send out their first draft. First drafts often contain errors that are embarrassing or that confuse the message. In your first draft, concentrate on *what* you are saying. In the second draft, concentrate on *how* you say it. These are different processes that should not be combined. Even professional writers find it difficult to edit for content while simultaneously editing for grammar, sentence structure, and spelling.

Writing Better Reports

Writing reports is an important task, especially when customers will see them. Robert Maidment recommends the following steps for writing good reports:[6]

1. *Define the problem.* Before beginning to write a report, you should be able to finish the following sentence: "The purpose of this report is. . . " The problem statement for a report should be brief, to the point, descriptive, and accurate.

2. *Develop a workplan.* A workplan is a list of tasks to be completed and a projected date of completion for each task. A workplan helps keep the development of the report on schedule. An effective way to structure a workplan is to develop a table of contents for the report and then list the tasks that must be accomplished under each major heading. Finally, assign a projected completion date to each task (see Figure 8.12 for an example of a workplan).

3. *Gather relevant data.* This is the research step; it involves collecting all data pertaining to the problem in question. This might include searching through files, reading other reports, interviewing employees and/or customers, running tests, and any other action that will yield useful data.

4. *Process findings.* Information is simply data that have been converted into a useful form. Processing findings means converting the raw data collected

TITLE OF REPORT: Failure Rate of Circuit Boards	
Major Report Headings	**Completion Date**
• Problem Statement	
1. Collect a broad base of input	January 15
2. Write problem statement	January 15
• Background	
1. Research all pertinent files	January 17
2. Analyze and synthesize	January 18
3. Develop a background summary	January 18
• Conclusions	
1. Develop conclusions	January 20
2. Summarize conclusions	January 21
• Recommended Solution	
1. Select the best solution	January 22
2. Pilot test the solution chosen	January 23
3. Summarize the results and the rationale for recommending the solution	January 24

FIGURE 8.12 Report workplan.

in the previous step into information on which recommendations can be based. This involves both analysis and synthesis.

5. *Develop conclusions.* Having identified a problem, gathered all pertinent data relating to the problem, and analyzed and synthesized that data, you next draw conclusions. The conclusions explain, based on data collected, analyzed, and synthesized, what caused the problem. Conclusions should be based on hard facts, objective, and free of personal opinions or editorializing.

6. *Make recommendations.* This section contains the writer's recommendations for solving the problem. They should grow logically out of the conclusions. Arrange recommendations in order of priority. Whenever possible and appropriate, give options. Recommendations should be specific and detailed, and should indicate timeframes, the people responsible for carrying them out, costs, and any other pertinent information.

COMMUNICATING BY TELEPHONE

Effective telephone communication is a critical element of good customer service. So much of a company's business is conducted by telephone that good telephone skills should be considered a requisite skill for every employee. Employees need to understand the most common customer complaints about telephone service, the essentials of telephone etiquette, how to organize themselves so that the information they need is available for telephone calls, how to take proper telephone messages, and how to overcome the disadvantages of voice-mail systems.

Common Customer Complaints About Telephone Service

All companies are affected by the never-ending upward spiral of customer expectations. As technology advances and becomes more convenient, our expectations for convenience increase. Few things have had a greater influence on the convenience of verbal communication than telephone technology—both land-line and cellular. As a result, people have come to expect instant access and instant communication. High expectations with regard to verbal communication have given rise to several common complaints from customers about the telephone service of businesses they deal with.

- I cannot get access to a human being (endless-loop voice-mail systems).
- There are too many rings; slow answering or no answer.
- The person who answers the telephone has a negative or unhelpful attitude.
- The message on the voice-mail system is of poor quality (I cannot understand what it says).
- I am left on hold too long.
- I am put through improper transfers.
- The person who answers the phone uses poor grammar and gives evasive or inconclusive answers.

Endless-loop, inconvenient, or overly complex voice-mail systems are a management issue in engineering, manufacturing, and construction companies. Employees should be sensitive to how customers respond to their company's voice-mail system and should provide continual feedback about it to managers. All employees should be trained to answer their telephones on the first or second ring. After more than two rings, customers feel that they are interrupting you, which may be the case, but that is not the message you want to send.

Employees also should be trained to be positive and helpful to customers who call, even when the issue is someone else's to deal with. It is critical that the customer be made to feel important and be ensured that he will be taken care of promptly. Do not just transfer a customer to the right person; ease his way by explaining whom you are transferring him to. Also, brief that person quickly about the customer's issue before connecting him.

People often make a bad impression with the outgoing messages on their voice-mail systems. This message may be the only information a customer has when forming an opinion of you and your company. Therefore, keep your recorded messages short, positive, and helpful. Speak slowly so that your name and any other information you provide are easily understood. Do not be flippant or cute. Save the cute or humorous messages for your home answering machine.

People do not like to wait. On the telephone, waiting periods are magnified psychologically. If it is necessary to keep someone on hold for more than a few moments, do one of two things: (1) keep checking in with the customer every 15 seconds or less so that she knows she has not been forgotten or cut off, or (2) offer to call the customer back rather than keeping her on hold. Being transferred to the wrong person or being transferred to someone who is out of the office can be irritating. When transferring a customer, get the intended individual on the line first, tell him you are about to transfer a customer, and briefly explain the customer's issue. If the intended person is not in, apply one of the following strategies: (1) find another individual who can help, and facilitate the transfer; (2) tell the customer you will locate the person and have him return the call as soon as possible; or (3) ask the customer if she would like to leave a message either with you (on paper) or on voice mail.

Your telephone voice should project an attitude of competence and professionalism. One of the worst telephone errors is the use of poor grammar or vague, inconclusive answers. Employees who sound ill-informed or poorly educated project the worst possible image for their company. Employees should be trained to use a professional voice and bearing on the telephone, and to save the vernacular for their close friends outside of the workplace.

It is perfectly acceptable to say, "I don't have that information at the moment, but I can get it for you right away." It is certainly preferable to giving vague, inconclusive answers that leave the customer hanging. Trust is an important element of customer service; customers who receive vague or inconclusive answers will most likely trust neither your company nor its employees.

Essentials of Telephone Etiquette

Good telephone etiquette is good business. In most businesses, the telephone rings hundreds or even thousands of times a day. It is a lifeline between companies and their customers, suppliers, bankers, and other

Positive Telephone Strategies

- Be businesslike and professional.
- Smile while talking.
- Practice common responses.
- Speak slowly and clearly.
- Use a positive tone of voice.
- Begin by identifying yourself.
- Answer promptly.
- Get the customer's name, and use it.
- Concentrate and listen assertively.

FIGURE 8.13 What to do when receiving a telephone call.

connections in the outside world. Every time the telephone rings, the image of your company is on the line. Incompetence or poor etiquette on the telephone can plant seeds of doubt in the minds of even the most loyal customers. Consequently, telephone etiquette should be viewed as a fundamental job requirement for all employees. This section explains both positive telephone strategies and telephone behaviors to avoid (see Figures 8.13 and 8.14).

Positive Telephone Strategies

The positive telephone strategies listed in Figure 8.13 are explained in the following paragraphs.

- *Be businesslike and professional.* Remember that your telephone presence should project an image of competence and professionalism. Be pleasant, courteous, positive, and businesslike.
- *Smile while talking.* There is a customer-service adage that says, "A smile can be heard over the telephone." Smiling while talking on the telephone puts you in a more positive, friendly frame of mind.
- *Practice frequently required responses.* If you are not comfortable on the telephone, practice commonly used conventions (*e.g.*, what to say when answering the telephone, what to say when transferring a caller to someone else, what to say when taking a message).

Telephone Behaviors to Avoid

- Doing paperwork at the same time
- Eating or drinking
- Talking with someone else in your office
- Interrupting or contradicting the caller
- Sounding rushed, bored, or annoyed
- Giving out unverified information
- Lying
- Sounding unsure or inconclusive
- Leaving customers on hold for more than 15 seconds

FIGURE 8.14 What *not* to do when using the telephone.

▪ *Speak more slowly than usual and speak clearly.* When speaking with someone in person, there are nonverbal cues that help you put your message across. When talking on the telephone, there is only the voice to convey the message. This fact, plus the natural human tendency to speak faster when on the telephone, can make it difficult for the customer to understand you. When talking on the telephone, make a point of speaking slowly and enunciating clearly.

▪ *Use a positive tone of voice.* Your tone of voice can be even more important than the words you use in a telephone conversation. Tone of voice has greater significance when talking on the telephone than when speaking with someone face-to-face. This is because the lack of other nonverbal cues magnifies the effect of voice tone. Your tone, when talking on the telephone, should be positive and upbeat. A bored, monotone, or hurried tone conveys to the customer a lack of interest on your part.

▪ *Answer the telephone with your name.* The best way to answer the telephone is with your name (*e.g.*, "Good morning, this is Melanie Bolter."). Some companies ask employees to give the company name as well. Although this is good advice for the switchboard operator, it is better for other employees to be brief and to the point when answering an office or cellular telephone. If a customer has gotten that far, she probably knows the company's name.

▪ *Answer promptly.* In this world of instant everything, a wait of more than two rings seems too long. People have come to expect an answering machine to pick up after two rings (or three at most). Too many rings can send a multitude of messages to a customer, all bad: (1) you are too busy to answer, (2) your office is disorganized and doesn't handle telephone calls well, (3) you do not care about him enough to answer promptly, or (4) you leave your telephone unattended.

■ *Get the customer's name at the outset, and use it.* People like to hear their name used. Even the most cynical, difficult customer is susceptible to the pleasant sound of his own name. Even if you have to ask for it, get the customer's name, and then use it immediately. Work it into the conversation several more times, being sure to pronounce it correctly. We recommend using the customer's name at least three times in a conversation that is three minutes or more in duration—that is, use the customer's name at least once a minute. Few techniques put a conversation on a positive, friendly footing faster than this one.

■ *Concentrate and listen assertively.* On the telephone you have only the customer's words and voice to go by. Avoid distractions, pay attention, and apply the assertive listening strategies explained earlier. Concentrate on both what is said and what is not said.

Telephone Behaviors to Avoid

Telephone behaviors to avoid are listed in Figure 8.14 and explained in the following paragraphs.

■ *Allowing distractions.* Distractions such as doing paperwork or talking with someone else nearby divert your attention during telephone conversations. Eating or drinking while trying to talk on the telephone is rude first of all, but also distracts you—and the customer. Distractions can cause you to misunderstand the customer's message or miss it altogether. When talking with a customer, block out or eliminate all distractions and devote your full attention to the call. Listen assertively.

■ *Interrupting or contradicting.* Interrupting or contradicting a customer only serves to make him angry or frustrated. If he is already in a negative frame of mind, interrupting or contradicting will just make matters worse. Remember your assertive listening strategies and use them.

■ *Sounding rushed, bored, or annoyed.* Customers want you to share their urgency regarding their problem or the information they need. If you sound rushed, bored, or annoyed, the customer will likely interpret your tone to mean, "She doesn't care about me and my problem."

■ *Giving out incomplete or unverified information.* One of the most frequent telephone errors is giving out incomplete or unverified information just to get the customer off the line. If you truly do not have time at the moment to find the information a customer needs, it is better to set a specific time to call her back than to give out incomplete or potentially inaccurate information. Providing questionable information might get the customer off the line, but this misguided strategy eventually creates another, even worse problem when the customer acts on the inaccurate information.

■ *Lying.* It is difficult and time consuming to win customers' trust. Consequently, few things are worse than lying to them. Nothing strains the customer–supplier relationship more than lying. Never lie to a customer, period. If you do not know how to answer a customer's question, say, "I don't have that information at the moment, but I will get it for you right away." If you promised to do something for the customer but forgot, admit the oversight, apologize, and do it now.

■ *Sounding unsure or inconclusive.* Part of the job of all employees in engineering, manufacturing, and construction companies is to project an image of competence and professionalism on behalf of their company. Customers want to know they can trust you and your company to competently, efficiently, and effectively meet their needs. Unsure or inconclusive responses on the telephone project a wishy-washy image. It is better to say conclusively, "I don't know, but I will find out right away" than to say, "I'm not sure."

■ *Leaving a customer on hold for too long.* People do not like to wait. People who are put on hold have a tendency to wonder if they have been cut off or forgotten about. Both happen often enough that this is not unrealistic. The only time a customer wants to remain on hold is when his issue needs immediate attention and he does not trust you to call him back promptly. If a customer indicates a willingness to be put on hold, check back with him at least once every 15 seconds or, better yet, add a feature to your telephone system that does this for you.

Organize Yourself for Better Telephone Service

Providing effective customer service by telephone is challenging under even the best of circumstances. If you are disorganized and your office is cluttered, it can be next to impossible. When customers call, you need to locate quickly the information they need and you must focus on their requests without the distractions of office clutter. The following strategies can help you get organized for better customer service over the telephone:

■ *Organize your work.* Some people keep their work neatly filed in folders, and pull out only the folders they need at a given time. Other people are visually oriented and like to see all their work, current and pending, arrayed on the desk before them. Regardless of which approach works better for you, the point is to be able to locate quickly the information you need when a customer calls. If you keep your work in files, keep the files nearby and know your filing system. If you keep your work in piles rather than files, go through the piles at the beginning and end of each workday, so that you know what is in each pile. It is important to have a system. Sort your piles by customer name, project number, priority order, or some other logical arrangement, and be familiar with what is in each pile. Never keep a customer waiting on the telephone because you cannot locate the information she needs.

- *Refile customer information immediately.* It is important to refile customer information immediately after completing a telephone call. It is easy to put aside the paperwork or drawings you were working with during the call and move on to your next task. It you do this just a couple times each day, before long you will have misplaced valuable customer information. Then, the next time the customer calls, you won't be able to locate the information. Even with the best intentions, you will sometimes forget or get too busy to refile a customer's paperwork. For this reason, we recommend using the last 15 minutes of each workday to refile paperwork and to get the files and piles you used today in order for tomorrow.
- *Keep customer files within reach.* Filing systems that require you to run up the hall to a central file room prevent ready access to information. Online storage has alleviated this problem somewhat, but it is not always completely implemented. Some companies still store their hardcopy files—particularly drawings—centrally. If this is the case with your company, make sure there are several telephones in the central file room, and to the extent possible keep copies of active customer files in your office.
- *Schedule call-return times on your daily calendar.* It is easy to get so caught up in the daily exigencies of work that you put off the task of returning telephone calls until the workday is almost over. This is a common problem, not just in engineering, manufacturing, and construction, but in every business. To counteract the problem, schedule time on your calendar every day for returning calls. Block out the time and guard it in the same way you guard the time you allot for appointments.
- *Make notes before calling customers.* What do you want to tell the customer? What do you want to ask the customer? What do you want the customer to do? Writing down the answers to these questions ahead of time can improve the efficiency and results of the call.

Taking Telephone Messages for Others

An additional telephone skill relates to taking a message for someone else. It is important to provide complete information. When taking a message for a colleague, apply the following strategies:

- Give the name and telephone number (including the area code) of the person who called.
- Record the date and time the call was received. Also ask the customer for and record the times he will be available to receive a return call. This will prevent your colleague from wasting time playing "telephone tag."

- Include your name on the message, in case your colleague needs to ask for additional information before returning the call.
- Summarize briefly what the customer needs.
- Give an indication of the customer's frame of mind (angry, worried, frustrated, rushed, etc.).
- Summarize briefly what you told the customer.

A detailed telephone message containing this information allows your colleague to return the telephone call equipped to provide good customer service.

COMMUNICATING BY EMAIL

Electronic mail has made it possible to have instant access 24 hours a day, 7 days a week, 365 days a year. On the other hand, it has magnified the problem of information overload. It has given people a way to convey information even faster than through express mail and facsimile machines. But on the other hand, it tends to reinforce poor planning and last-minute submittals. It allows people to send messages without the problems and delays associated with telephone calls. But some argue that email is diminishing the ability of users to write clearly.

Technology is always a mixed blessing, and this is certainly true of email. What follows are several strategies for ensuring that email is a positive communication vehicle for you and your company.

- *Remember that the recipient is a human being.* With face-to-face communication, the physical presence of the recipient has a restraining effect on what you say and how you say it. This is not the case with email. Consequently, you may find yourself saying things in an email message that you probably would not say in a face-to-face conversation or over the telephone. It can be easy to forget that there is an actual human being at the other end of an email message. To counteract this tendency, force yourself to think of the individual who will be receiving the message.
- *Consider the recipient's time.* With email, sending large documents is as easy as clicking on an icon. However, easy to send does not necessarily mean easy to receive. Depending on the capabilities of the recipient's equipment, downloading and printing large documents can be a time-consuming task. Consequently, never send large, complicated documents, especially those containing high-density graphics, without the prior approval of the recipient. It is poor customer service to tie up a customer's computer and printer for an extended period of time with a document that could have been sent by express mail.
- *Remember grammar, spelling, punctuation, and sentence structure.* Email has rapidly become to written communication what informal, hurried talk

is to verbal communication. Too often, people who use email drop all pretenses of proper grammar, spelling, punctuation, and sentence structure. It is as if they think the inadequacies of their writing somehow won't be noticed. Unfortunately, they will. When writing email messages, remember that they affect your company's image as much as any other form of written communication does.

■ *Cool off before you write.* The *instant-communication* characteristic of email can be a blessing and a curse. Abraham Lincoln used to give this advice: If you are angry with someone, sit down and write the offending party a blistering letter telling him exactly what you think. Then put the letter aside for at least a day. After a day has passed, pick up the letter and reread it. Chances are, having cooled off overnight, you will tear the letter into little pieces, and throw it away. Unfortunately, many people who use email don't take Lincoln's advice. This is when the ability to send an instant message can be a curse. Click on "Send," and a message you'll later regret is on its way to a recipient who will not appreciate receiving it. You'd do well to remember Lincoln's advice as it might be recast for modern times: *Never write or send an email when you are angry.* This advice is even more important when the recipient is a customer.

DEVELOPING INTERPERSONAL SKILLS

Interpersonal skills enable people to work together in a manner conducive to both personal and corporate success. Functioning effectively in an ECS setting requires good interpersonal skills. Positive interpersonal relations are of critical importance among team members, between company representatives and customers, among internal customers, and between company officials and vendors.

The following strategies lead to building a workforce with good interpersonal skills:

■ *Recognition of the need.* Managers must recognize the need for interpersonal skills in their employees. Historically, the recruitment in engineering, manufacturing, and construction companies has focused on technical skills, degrees, diplomas, and so on. These are certainly important considerations in staffing decisions, but to them must be added interpersonal skills.

■ *Careful selection.* When interpersonal skills are included as a criterion in the selection process, the process changes somewhat. The screening of written credentials and technical skills continues in the normal manner. After the candidates with the best credentials and technical skills have been identified, they are then carefully screened to determine whether they have essential interpersonal skills—listening, patience, empathy, tact, open-mindedness, friendliness, and the ability to get along and be positive agents in a diverse workplace.

- *Training.* It is the uncommon individual who possesses inborn interpersonal skills. Some people are naturally good at relating with others. Most of us, however, have room for improvement in this area. Fortunately, interpersonal skills can be learned. People can learn to listen better, empathize with different types of people, be tactful, and facilitate positive interaction among fellow employees. Consequently, companies should add the topic of interpersonal skills to their employee training programs.
- *Measurement and reward.* If managers value interpersonal skills, they will measure these skills as part of the normal performance appraisal process. Correspondingly, the results of such appraisals will be built into the reward system.

Human Connections in the World of Technology

Consider the many communication-enhancing technologies that are a normal part of modern life. The television, radio, telephone, facsimile machine, Internet, and cellular telephone were all developed in the name of improved communication, but have they *really* improved communication?

In some ways, technology has actually had a negative effect on communication, because it has removed much of the need for face-to-face interaction. The key to communication is perception. Receivers must be able to perceive feelings, emotions, intent, and other intangible, nonverbal aspects of a message that are missing in a facsimile transmittal, an email, or even a telephone conversation. For example, in certain situations a simple handshake can communicate a powerful message—a message that cannot be transmitted by technology.

Human connections have become more important than ever because technology has made it so easy to interact without really communicating. Engineering, manufacturing, and construction companies should be especially attentive to human interrelationships, interdependence, and interaction and to the negative impact technology can have in these areas. The next section explains what companies can do to promote human connections in today's technological workplace.

Promoting Human Connections in a Technological World

The goal in promoting human connections is responsiveness. Responsive people can perceive the real message from among the verbal, written, and nonverbal cues they receive. Responsive employees and managers are assets in an ECS setting; they are more likely to be effective at delivering customer service than unresponsive employees are.

Managers can promote responsiveness in the workplace by encouraging employees to apply the following strategies:

■ *Respect people.* Remember that people—both customers and fellow employees—are the most important resource in any business. Managers and employees who value people treat all people with respect, not just those who have something they want.

■ *Give people what you want to get back.* People have a natural tendency to mirror the treatment they receive from others. Typically, people who treat others with decency and respect are, in turn, treated with decency and respect. People who are loyal to others are likely to receive loyalty in return. Encouraging employees to get by giving is a worthwhile undertaking for companies in a competitive marketplace.

■ *Make cooperation a habit.* World-class athletes practice the skills that make them great until they become automatic and habitual. Any habit is hard to break. Just ask people who have tried to stop smoking or biting their nails. Employees who practice cooperation until it becomes habitual will practice cooperation for life.

Cooperation in the workplace means learning to use the word "we" instead of "I" and "they." It means chipping in to get the job done even when it is not part of one's own job description. It means involving all employees who must do the work in decisions regarding that work.

Cooperation does not mean always saying "yes." When the right answer to give is "no," that answer should be given without a sugar coating. Cooperation in such cases means showing people they are valued by explaining why the answer must be "no."

Summary

1. *Communication* is the transfer of a message that is both received and understood. *Effective communication* is a higher order of communication. It means the message is received, understood, and acted on in the desired manner.

2. Communication is a process that involves a message, sender, receiver, and medium. The message is what is transmitted (information, emotion, intent, or something else). The sender is the originator of the message, and the receiver is the person for whom the message is intended. The medium is the vehicle used to transfer the message.

3. Various factors can inhibit communication. The most prominent inhibitors are differences in meaning, lack of trust, information overload, interference, premature judgments, the kill-the-messenger syndrome, a condescending tone, inaccurate assumptions, and listening problems.

4. A climate conducive to communication results in people receiving the information they need to do their jobs, it builds morale, and it promotes creativity. A negative communication climate creates conflict, confusion, and cynicism.

5. One of the most important communication skills is listening. Good listening means receiving the message correctly, decoding it, and accurately perceiving what it means. Inhibitors to good listening include lack of concentration, preconceived ideas, thinking ahead, interruptions, tuning out, and interference.

6. Listening skills can be improved by upgrading the desire to listen, asking the right questions, judging what is really being said, and eliminating listening errors. Body factors and proximity must also be managed carefully to maximize listening skills.

7. Verbal communication can be improved by showing interest, being friendly, being flexible, being tactful, being courteous, dropping your defenses, stating your purpose, acknowledging emotions, and using carefully phrased open-ended questions.

8. Written communication can be improved by being brief, being direct, being accurate, and practicing self-editing. The following step-by-step strategy will help managers write better reports: (a) define the problem, (b) develop a workplan, (c) gather relevant data, (d) process findings, (e) develop conclusions, and (f) make recommendations.

9. The most common complaints customers make about telephone service are: (a) inability to get access to a human being; (b) too many rings—slow answering; (c) negative unhelpful attitudes; (d) poor-quality voice messages; (e) being left on hold for too long; (f) improper transfers; and (g) poor grammar.

10. Interpersonal skills enable people to work together in a positive manner conducive to both personal and corporate success. To ensure that employees have good interpersonal skills, managers should recognize the need for interpersonal skills, select personnel carefully, provide training, measure the skills, and reward them.

Key Phrases and Concepts

Body factors
Communication
Company-level communication
Condescending tone
Effective communication
Good listening
Inaccurate assumptions
Information overload
Interference
Interpersonal relations
Lack of concentration
Medium
Message
One-on-one communication
Open-ended questions
Preconceived ideas
Premature judgments
Proximity

Self-editing
Sender
Team-level communication
Thinking ahead
Tuning out
Verbal communication
Voice factors
Written communication

Review Questions

1. Define "communication" and "effective communication."
2. What are the four levels of communication?
3. Explain the process of communication.
4. List and briefly explain six inhibitors of communication.
5. List and briefly explain five inhibitors of good listening.
6. Explain four strategies for improving listening skills.
7. What are the following factors, and how do they affect listening: body factors, voice factors, proximity factors?
8. How can a person improve her verbal communication skills?
9. List and explain five rules of thumb for improving written communication.
10. What are the steps for improving written reports?
11. Six guidelines to improved communication were set forth in this chapter. Explain all six.
12. How can companies ensure that their employees have good interpersonal skills?

ECS APPLIED: DIVERSIFIED TECHNOLOGIES COMPANY IMPROVES COMMUNICATION

In the last installment of this case, Jake Arthur, DTC's vice president for construction, gave a report to the other vice presidents and the CEO on the progress in helping employees in his division learn to deal with dissatisfied and difficult customers. He shared a success story about one of his project superintendents who had a well-deserved reputation for being a grouch. Surprisingly, the training this employee had received in dealing with difficult customers seemed to be working. Meg Stanfield and Tim Wang also reported that their divisions had experienced equally encouraging results.

"Let's review where we are in the ECS implementation," said David Stanley, CEO of DTC. "Effective communication is essential to everything we are trying to do, so I'm anxious to hear about our progress. Have all of you tested our telephone system?" The CEO had asked each of his vice presidents to call various individuals in their divisions from an offsite telephone to test the responsiveness of the company's telephone system. He had also asked them to call several long-time customers to solicit their feedback concerning the system's responsiveness.

"I'll get us started," offered Meg Stanfield, vice president for engineering. "I discovered several annoying glitches when testing the system." Stanfield explained that she had called the company's main number as if she didn't know the direct access number for one of her engineers, and had found the automated answering service to be both confusing and frustrating. When the system asked her to key in the last name of the individual she wished to speak with, it cut her off after just three letters and forwarded the call to a different person, one with a similar last name.

The system did this three times in a row. Stanfield also expressed concern that it took a long time to listen to all the various automated options. Her calls to long-time customers had verified her concerns. Tim Wang and Jake Arthur agreed with Stanfield and explained that they had experienced similar problems. Seeing Stanley's concern, the three vice presidents assured the CEO that they had already met with the company's telephone specialist, who was working on improvements right then.

Jake Arthur looked at the other two vice presidents and asked, "Who wants to tell David about the telephone message issue?" "You do it, Jake," said Wang and Stanfield in unison. Turning to the CEO, Arthur explained, "Our customers really don't like to leave messages with our personnel. Apparently, a lot of customers have left messages only to have them either not passed on or passed on inaccurately. One customer told me he had left a message for Linda Sheffield about the new hospital we are building over on Mulholland Drive. Linda is a project manager in the construction division. The call was eventually returned by Melinda Colfield, who thought it was the hospital calling about her daughter who had just gone through surgery."

"Apparently, we do all right when customers leave voice-mail messages, but when it comes to messages left with one employee for another, we have problems." Arthur then told Stanley that as part of the strategy for solving this problem, the three vice presidents had developed a standardized telephone message form for companywide use. He also explained that all employees were going to receive training in the effective and proper use of the telephone. "Here is the good part," said Arthur with a smile. "David, you are going to be a movie star." Seeing the puzzled look on the CEO's face, Arthur went on to explain. "Our public relations department has arranged to produce a training video for DTC on telephone etiquette. One section in the video will deal with taking telephone messages. You are going to do the introduction at the beginning of the video and narrate the section that talks about our new companywide telephone message form. After you show the new form and explain that all employees will be expected to use it, we are going to have two cases that show what can happen when telephone messages are taken poorly."

"Good idea," acknowledged Stanley. "I'll be happy to work with the production team on the video." Stanley then told the vice presidents several war stories about bad experiences with inaccurate telephone messages. Not to be outdone, the vice presidents shared several stories of their own. Before dismissing the meeting, Stanley asked the vice presidents to discuss their plans for the next step in the implementation—establishing internal customer satisfaction.

DISCUSSION CASES

The following cases provide examples of how the various concepts presented in this chapter might play out in actual companies. The cases are provided to prompt discussion, give the reader a feel for the types of problems confronted in the workplace, and reinforce the ECS concept in question.

CASE 8.1 Poor Communication with International Customers

Marcum Construction Company was rapidly becoming Marcum Construction *International*. For the past two years, most of the company's new business had been in South America, Central America, and the Caribbean. International expansion presented Marcum Construction with both opportunities and problems. On the one hand, with international expansion Marcum's opportunities for growth were almost limitless. On the other hand, company officials found themselves increasingly bogged down in a mire of new and different problems.

After studying the situation, CEO Josh Marcum felt sure that poor communication on the international projects was at the heart of his company's new problems. Specifically, Marcum found a major lack of trust and differing perceptions between his project managers for each international contract and the local people hired to do the work. Even though all Marcum's project managers on international contracts spoke fluent Spanish, they were born and raised in the United States and did not speak the local dialects of the communities in which their construction projects were located. Consequently, they were viewed as outsiders. This was an especially difficult problem because winning the trust of local people could take years, and Marcum did not have years. All of his international contracts must be completed in 24 months or less.

Then there was also the problem of differing perceptions. Project managers born, raised, and educated in the United States have radically different perceptions about work and how best to get it done than do locals from countries in the Caribbean, South America, and Central America. As one of his project managers told him, "It's a whole different world down here."

Marcum struggled for several weeks trying to decide what to do about his company's communications problems. He eventually decided he needed the help of an expert. Once the decision to seek help was made, he took bold action. Marcum called the United States Department of Commerce in Washington, D.C., and scheduled an appointment to meet with an expert on doing business in Central America, South America, and the Caribbean. The advice he received was simple: Hire a local expediter for each project. Make sure the individual is known, respected, and trusted in the local community. Also make sure he knows how the local government works. Use the expediter as a go-between when hiring, firing, and supervising workers. Finally, tie the expediter's compensation to specific completion points for the project.

Marcum, before talking to the Commerce Department official, would have viewed hiring local expediters as an unnecessary and unwarranted expense. Within just weeks, with local expediters in place, the communication problems began to go away. And Marcum's projects were progressing on schedule. This experience changed Marcum from a skeptic to a believer when it comes to the importance of communication. In fact, he is now such an advocate of effective communication that he hires local expediters not just on international projects, but also on some of his domestic jobs.

Discussion Questions

1. Should Marcum have predicted the communications problems his company had in this case? Explain.
2. Analyze how Marcum handled this situation. Would you have done anything differently?

CASE 8.2 Our Writing Is Terrible!

Mary Audrey was embarrassed. As CEO of Audrey, Gleason, and Boyd Engineers, she took pride in the quality of her company's work. In fact, the corporate motto of A, G, & B was "We strive for perfection." The errors and other shortcomings in the letters she was holding in her hand violated the corporate motto. That fact was clear enough. But to make matters even worse, this embarrassing fact had been pointed out to her by a customer—and not just any customer, but her former college roommate.

The former roommate, Louan Roman, was also an engineer and CEO of a company she founded. Roman was an aerospace engineer, whereas Audrey's field was structural engineering. For the past year, Audrey's company had been building a new manufacturing facility for Roman's company. Now that the ribbon had been cut and Roman's employees had moved in, Audrey and Roman were having lunch to celebrate.

"Do you remember how I hated to write when we were in college?" asked Roman. "I sure do," answered Audrey, knowing where the conversation was headed. "When I would criticize your writing, which was terrible, you would always say the same thing. Do you remember what you would say?" asked Audrey. "Yes. I would say I'm going to be an engineer, not a novelist." "That's right," confirmed Audrey. "And how did I respond to this half-baked excuse?" Roman made a show of attempting to search her memory. Then she smiled at her friend. "You'd say, real engineers can write."

"I know. That's why I'm so embarrassed about these letters," said Audrey with a sigh. "They read like the work of a sixth-grader." "They aren't that bad," replied Roman seeing her friend's discomfort. "Don't be embarrassed. I just gave them to you because I know from personal experience that you expect better work than this from your employees. Since you wouldn't let me get away with sloppy writing in college, I knew you wouldn't approve of your employees sending out poorly written letters." "I don't approve of sloppy work," said Audrey emphatically. "It's just that my company has gotten so big that I no longer see everything that goes out. When you and I were just starting our companies, I read every piece of correspondence that was sent out on A, G, & B stationery. Now I have more than 100 personnel writing letters and reports every day." "Same here," said Roman with a knowing nod. "Sometimes I get nostalgic for the good old days when we were just a couple of rookies with big ambitions and bigger dreams." "Me too," replied Audrey with a wistful look. "In fact, I feel that way right now."

"Mary, why don't you give your employees that same writing seminar you gave me back in college?" suggested Roman. When Audrey and Roman were in college, Roman had struggled with any class that required writing—so much so that her grade point average had suffered. To help her, Audrey—who was an excellent writer—had taken Roman to the university's writing lab one Saturday and worked with her all day. They had planned, written, and critiqued letter after letter and essay after essay. By suppertime, both students were exhausted, but happy. Roman's writing had improved significantly, but what was even more important was that she no longer dreaded writing. She had started the day thinking people who wrote well just sat down and started writing. Roman did not realize that good writers planned, outlined, used tools such as dictionaries and thesauruses, wrote drafts, and performed self-editing. Once Audrey had shown her how to plan first and write second, Roman's work had gone from poor to good in less than an hour.

"That's exactly what I need to do," replied Audrey. "You are living proof that even the most reluctant engineer can learn to write well." "Yes we can," said Roman with a smile. "I am going to organize a seminar for my employees right now and make it mandatory," concluded Audrey, clearly feeling better now that she had a plan of action. "Thanks for bringing me these letters, Louan, and thanks for the advice." "No, thank you," said Louan Roman. "What for?" asked a puzzled Mary Audrey. "First, for teaching me to write so I could graduate from college. Second, for building me a wonderful new plant. Your people did an excellent job."

Discussion Questions

1. Do you consider yourself a good, average, or weak writer?
2. Do you think poor writing affects a company's image? Explain.

CASE 8.3 The Telephone Is Hurting Us

"The reason I called this meeting today is to talk with you about the telephone," said John Kirby, plant manager for CoopTech Manufacturing. Kirby's statement took his managers by surprise. They thought the meeting would be about the usual concerns: production schedules, process improvements, machine maintenance, and reject rates. Kirby continued, "The telephone is hurting us." "It sure is," interjected Demontry Jones, Shift A supervisor. "Why don't we just unplug our hardwired phones and turn off our cellular phones? Maybe then we could get some work done around here." Kirby noticed that several other supervisors nodded in agreement.

"We can talk about limiting incoming calls during peak production hours if we need to," replied Kirby. "But right now I want to talk about taking telephone messages." "That's for secretaries to worry about John," snapped Jones, clearly irritated about the topic. "We've got processes to run and deadlines to meet." "I know, Demontry. I didn't start out here as plant manager. I was Shift A supervisor when you were in high school," retorted Kirby. Kirby had started his career at CoopTech as an industrial-engineer-in-training right after graduating from college. During his 15 years with the company, he had held every key position in the manufacturing division, including the critical job now held by his friend and colleague, Demontry Jones.

Sensing Kirby's frustration, Jones apologized. "I'm sorry, John. It's just that the telephone on the plant floor has been ringing off the hook lately, and all three shifts are behind schedule." "It's alright Demontry. I apologize too. I'm a little frustrated right now." "What's this really about John?" asked Beverly Chong-Andrews. Kirby took a deep breath to regain his composure before responding. Then he said, "All of you know that I have been trying to convince the Department of Defense to let us use a different paint on the missile container project." "Of course," said Joe Perkins, foreman of the painting department. "That new X-Tech paint would reduce our application and drying time by 50 percent, not to mention that it's nontoxic." Several other supervisors nodded in agreement.

"Well folks," said Kirby. "I have good news and bad news. The good news is that after six months of fighting with the federal bureaucracy, I've finally convinced the Department of Defense to let us change to the X-Tech paint." "That's great!" shouted Joe Perkins, while others cheered. But Demontry Jones, sensing Kirby's frustration, interrupted the celebration and asked, "What's the bad news John?" "The bad news is that our Department of Defense contact claims he called here three days ago and left the message that we should convert to X-Tech paint immediately." "Who was the message for?" asked Joe Perkins. "You," said Kirby.

"Me?" asked Perkins, clearly surprised. "I don't remember getting a message from the Department of Defense, and believe me that's a message I'd remember. The only telephone message I got this week turned out to be a wrong number." "Do you mean this one," asked Kirby, holding up a crumpled scrap of paper. "That's it," replied Perkins. "Where'd you get that?" "I pulled it out of the trashcan next to the shop floor telephone this morning after I talked with our Department of Defense contact. I guess it's just luck that the janitors hadn't emptied that trashcan in three days, although it's obvious we have a janitorial problem, too. But that will wait for another time."

"How many missile containers have we painted in the last three days, Joe?" asked Kirby. "More than 200," responded Perkins, "Why?" "Because now that the X-Tech paint has been approved, we have to strip every one of them down to the metal and repaint them. Apparently, the minute the Department of Defense approved the new paint, they expected us to start using it. That was the purpose of the call you never got, Joe." The chorus of cheers heard earlier was now replaced with a series of groans. Stripping the old toxic paint off 200 containers was going to be time consuming, difficult, and hazardous.

Kirby then passed around the message he had retrieved from the trashcan. It read as follows: "Joe, call 210-459-6682." "For the sake of clarification," said Kirby, "our Department of Defense contact's telephone number is 210-459-6862." Kirby then handed out a telephone-message form to all participants and said, "From now on, we will use this form companywide for taking telephone messages." The form required the name of the caller, the date and time of the call, what the caller wanted, a good time to return the call, and the name of the person taking the message. Kirby asked all participants to meet with their direct reports to distribute the new telephone-message form and to emphasize that all employees should use it. When one of the participants grumbled that it would take too long to fill out such a form, Kirby said, "Which would you rather do: fill out this form, or strip and repaint 200 missile containers?" The grumbling stopped.

Discussion Questions

1. Have you ever experienced a problem similar to the one in this case? Explain.
2. Analyze how Kirby handled this problem. Would you have done anything differently? If so what?

Endnotes

[1]Corwin P. King, "Crummy Communication Climate (and How to Create It)," *Management for the 90s: A Special Report from Supervisory Management* (Saranac Lake, NY: American Management Association, 1991), p. 21.

[2]Ibid.

[3]Ibid.

[4]C. Glenn Pearce, "Doing Something about Your Listening Ability," *Management for the 90s: A Special Report from Supervisory Management* (Saranac Lake, NY: American Management Association, 1991), pp. 1–6.

[5]Roger Ailes, "The Seven-Second Solution," *Management Digest* (January 1990): 3.

[6]Robert Maidment, "Seven Steps to Better Reports," *Management for the 90s: A Special Report from Supervisory Management* (Saranac Lake, NY: American Management Association, 1991), pp. 11–14.

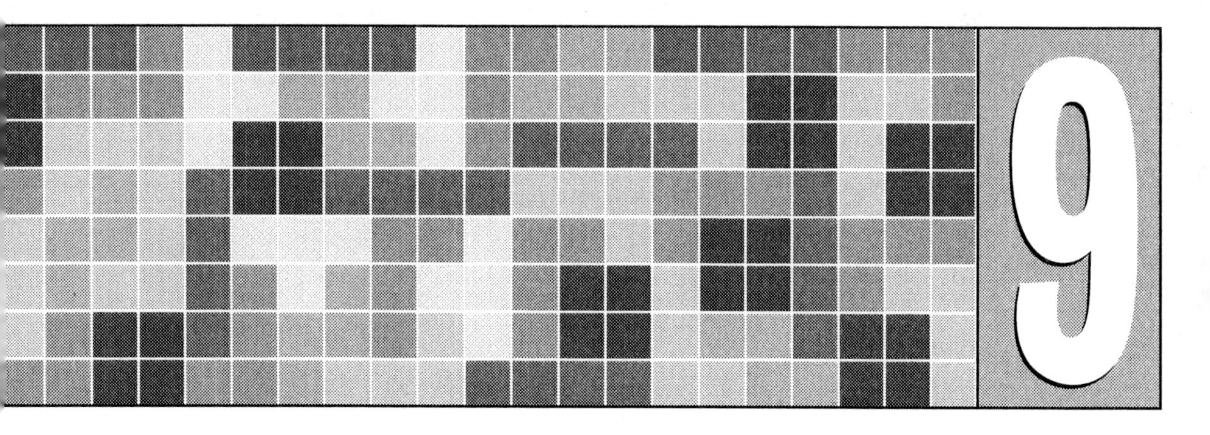

Establish Internal Customer Satisfaction

If you are an executive, manager, or supervisor who wants employees to treat customers well, remember this: Employees typically treat customers the way you treat employees.

GOALS

- Understand why internal customer satisfaction is important.
- Describe how leadership differs from management.
- Learn how to communicate effectively with employees.
- Explain why listening to employees can improve performance.
- Name ways to improve the job satisfaction of employees.
- Explain the concept of employee empowerment.

We tend to think of customers only as the entities that purchase and use our products—and some of them are. They are *external customers*, and all discussion of customer service in this book thus far has pertained specifically to this constituent group. As important as external customers are—and they are critically important—it is essential to understand that there are also *internal customers*. Internal customers are employees who depend on each other to get their work done. Part of satisfying external customers is tending to the morale, attitudes, and perspectives of the employees who interact with them. In other words, part of ensuring external customer satisfaction is ensuring internal customer satisfaction.

WHY INTERNAL CUSTOMER SATISFACTION IS IMPORTANT

Employees tend to treat customers the way they—the employees—are treated by management. Therefore, internal customer satisfaction must be a high priority. Low morale among employees can undermine a company's best efforts to ensure customer satisfaction. Employees are intuitively perceptive. They quickly notice any discrepancy between the company's expectations of them and its treatment of them. For example, employees will surely notice the following behaviors:

- Management expects employees to listen to customers, but it seldom listens to employees.
- Management expects employees to communicate frequently with customers, but it seldom communicates with employees.
- Management expects employees to seek customer input, but it seldom seeks employee input.
- Management expects employees to control their tempers when dealing with customers, but it allows supervisors to verbally abuse employees (or shoots the messenger when an employee points out a problem).
- Management expects employees to go the extra mile for customers, but it refuses to do so for employees.
- Management expects employees to apply ECS strategies when dealing with customers, but it refuses to do so itself (the "do as I say not as I do" model).

The rationale for establishing and maintaining internal customer satisfaction is this: In the long run, a company cannot maintain external customer satisfaction unless it also maintains internal customer satisfaction.

DO NOT MANAGE EMPLOYEES—LEAD THEM

If you want to improve the performance of your organization in the ECS area, or in any other area, *manage processes and lead people*. We define "leadership" in this context as follows:

The ability to inspire people to make a total and willing commitment to accomplishing organizational goals.

Implicit in this definition is the need for employees to make a commitment to something larger than themselves. Leaders inspire people to willingly commit themselves to accomplishing something that transcends their immediate personal interests. The larger commitment should be to the organization's goals.

Inspiring people in this way is a higher-order management skill than motivating them. Motivating people is a matter of persuasion, typically through wisely applied incentives. Motivation appeals to an individual's personal interests, not to a larger cause. Inspiring people, on the other hand, is a matter of inducing them to commit to a larger cause by providing an example they would like to emulate.

One of the best examples of the difference between motivation and inspiration can be found in America's war for independence from Great Britain. The British king hired thousands of German mercenaries, known as Hessians, to help fight on the British side. The Hessians cared little one way or the other about America's desire for freedom or Britain's desire to retain one of its most lucrative colonies. They were motivated by money. The American soldiers and minutemen, on the other hand, were fighting for a cause much greater than money; they were fighting for their freedom. Consequently, they were inspired by George Washington and other American revolutionary leaders to endure unimaginable hardships and to persevere against impossible odds. No amount of motivation in the form of money or persuasive words could have convinced the Americans to commit themselves to their cause as they did. Because they were inspired by their leaders and the desire for freedom they shared with their leaders, the colonists eventually prevailed.

Managers and supervisors who seek to inspire their employees to make a total and willing commitment to accomplishing organizational goals should apply the following strategies:

- *Set a positive example.* Expect employees to do as you do, not just as you say. Model the behaviors you expect of employees, and do so consistently. Inconsistent role-modeling is as ineffective as no model at all. Let employees see your expectations of them lived out consistently every day in your behavior.
- *Adopt a participative leadership style.* Participative leadership involves charting a course for the organization (department, team, and so on), and then involving employees in determining the best way to get to the desired destination. In an environment of participative leadership, people affected by decisions are empowered to provide input into the decision-making process. This does not mean that managers and supervisors give up their authority or responsibility. In point of fact, they can do neither, even if they want to. Rather, it means casting a wider net to collect input before making a decision. It is the opposite of imposing decisions on the people who must

carry them out without first soliciting their opinions. Participative leaders acknowledge that employees represent a vast resource for continually improving performance and treat them accordingly. Morale is enhanced when employees know they are empowered.

- *Have a sense of purpose.* People will not follow someone who appears aimless or lost. Decide exactly what the goal is (in the case of this book, customer satisfaction), and let employees know. People like to follow others who know where they are going, who communicate their purpose, and who demonstrate their commitment by example. Employee morale is enhanced when employees sense that managers and supervisors are committed to a worthy purpose.

- *Apply self-discipline.* Setting the right example day in, day out can be difficult. It is easy to become frustrated and give up. Everyone has a bad day occasionally; true leaders exercise self-discipline to control how they respond to adversity. Employee morale is enhanced when employees see managers and supervisors consciously modeling desired behaviors, especially in difficult circumstances.

- *Be honest with employees.* People will not follow someone they do not trust. It is better to tell employees an unwelcome truth than to tell them a sugar-coated falsehood. Employee morale is enhanced when employees know they can trust managers and supervisors.

- *Establish credibility with employees.* Employees will follow your example only if you have established credibility with them. This is accomplished by consistently doing yourself what you expect them to do. Employee morale is enhanced when they feel managers and supervisors are credible.

- *Demonstrate your commitment.* People are more likely to follow someone who is committed to an organizational purpose than someone who is not. Employees who can see from managers' behavior day in, day out that they are committed to customer satisfaction are more likely to make a similar commitment themselves. Morale is enhanced when managers and supervisors demonstrate their commitment to a higher purpose.

- *Be fair and impartial.* When monitoring employee performance as it relates to customer service (or any other area), it is important that managers and supervisors be fair and impartial. Showing favoritism dampens morale. Morale is enhanced when employees know they will be treated fairly and impartially.

- *Take the blame, but share the credit.* Managers and supervisors who engage in finger pointing and buck passing are misleaders, not leaders. On the other hand, managers and supervisors who accept responsibility for the performance and behavior of their direct reports inspire loyalty and a commitment to improvement in their team members. It is best for managers and supervisors to take the blame when it is publicly aimed at their direct reports, and then to settle the issue with the relevant employees in private. Correspondingly, when credit is due, managers and supervisors must be willing to share it with employees. Unselfishness on the part of managers and supervisors enhances morale.

COMMUNICATE WITH EMPLOYEES

Internal customer satisfaction is impossible without open, frequent, effective communication with employees. The same assertive listening strategies advocated elsewhere in this book should be used with employees. In addition, the following strategies should be applied:

- *Do not shoot the messenger.* Few things inhibit communication between employees and their supervisors more than the shoot-the-messenger syndrome. Managers who become irritated with and take out their irritation on an employee who brings them bad news will soon find that they receive no more bad news. This might sound appealing on the surface, but the problem with not receiving bad news is that if a problem is not dealt with promptly, it only gets worse. It is better to deal with a problem when it is still small than after it has grown to disastrous proportions. When there is a problem, managers and supervisors need to know about it whether they want to or not. Smart managers and supervisors thank employees for keeping them informed, even when the news is bad. Even smarter managers and supervisors encourage employees to recommend a solution when pointing out a problem.
- *Communicate corrective feedback constructively.* The differences between constructive criticism and just plain criticism are in the intent and the delivery. Criticism is constructive if it is intended to help the individual being criticized improve his performance. It is not constructive if it is intended or delivered in such a way as to punish, belittle, or embarrass the employee. The guidelines in Figure 9.1 can help ensure the effectiveness of corrective feedback.
- *Continually improve your communication with employees.* The following strategies can help improve your communication skills:
    - *Keep up to date.* Stay up to date with information relating to your workplace. You cannot communicate what you do not know.
    - *Prioritize and determine time constraints.* Communicating does not mean simply passing on to employees everything you learn. That would only overload them and inhibit communication. Analyze information and decide what your employees need to know. Then prioritize it according to urgency and share the information accordingly.
    - *Decide whom to inform.* Once the information has been prioritized, decide who needs to have it. Employees have enough to handle without having to process information they don't need. Don't

Guidelines for Communicating Corrective Feedback

- *Be positive.* To be corrective, feedback must be accepted and acted on by the employee. This is more likely to happen if it is delivered in a positive manner.
- *Be prepared.* Focus on facts. Do not discuss personality traits. Give specific examples of the behavior you would like to see corrected.
- *Be realistic.* Make sure the behaviors you want to change are controllable by the employee. Don't expect an employee to correct a behavior she does not control. Find something positive to say. Give the employee the necessary corrective feedback, but don't focus only on the negative. Tell the employee about the behavior, ask for her input, and listen carefully to her response.

FIGURE 9.1 Constructive criticism should lead to improved employee performance.

withhold information that employees *do* need, however. Achieving the right balance will improve communication.

- *Determine how to communicate.* Choose among the various ways to communicate (*e.g.*, verbally, electronically, one-on-one, in groups). A combination of methods is probably the most effective.
- *Communicate the information.* Don't just tell your employees what you want them to know. Ask questions to determine if they have really gotten the message. Encourage them to ask clarifying questions. Agree on the next steps they should take based on the information.
- *Check accuracy and get feedback.* Follow up to confirm that your communication was accurately received. Can the employees paraphrase and repeat your message? Are the employees undertaking the next steps as agreed? Get employee feedback to ensure that their understanding has not changed and that progress is being made.

- *Select the appropriate communication method.* Most workplace communication is either verbal or written; you need to know when each method is most effective. Verbal communication is most effective when used in the following situations:

- *To communicate a message requiring immediate action on the part of employees.* The more appropriate approach in such a case is to communicate the message verbally and follow up in writing.
- *To commend an employee for doing a good job.* This should be done verbally and publicly. Then follow up in writing.
- *To reprimand an employee for poor performance.* This message can be communicated more effectively if given verbally in private. This is particularly true for occasional offenses. Follow up in writing.
- *To resolve conflict between or among employees about work-related problems.* The necessary communication in such instances is most effectively given verbally and in private.

Written communication is most effective when used in the following situations:

- *To communicate a message requiring future action* on the part of employees.
- *To communicate general information,* such as company policies, personnel information, directives, or orders.
- *To communicate work progress* to an immediate supervisor or a higher manager.
- *To promote a safety campaign.*

■ *Use communication to motivate.* Just as people have a natural instinct for competition, they also have a natural desire to be informed. They want to know how they are doing as individual employees, and how the team, the division, and the company are doing. Providing up-to-date, accurate information on a continual basis is an excellent way to motivate employees. In fact, without effective communication, all the other motivation strategies break down.

Communication involves more than just conveying information. It also involves giving instructions, listening, persuading, inspiring, and understanding. Communication can be viewed as preventive maintenance. To use communication to prevent problems and to keep small problems from becoming big ones, you must tune into what is being said by employees, verbally and nonverbally, as well as what is going unsaid. This requires that you have the ability to:

- understand nonverbal communication
- empathize with employees and see things through their eyes
- read between the lines and determine what the real problem is
- keep an open mind and truly listen to what employees are saying

■ *Use electronic communication wisely.* Email enables frequent contact with employees. This is good, provided the technology is used wisely. Email is so easy and convenient to use, however, that the temptation is to let it replace all other forms of communication. This would be a mistake. One-on-one conferences, informal talks in the hallway, team meetings, company newsletters, telephone calls, banners, and other communication methods are also very useful—and can be more appropriate. You should also consider that variety and repetition of a message are important when communicating with employees. They may need to hear the message several times and in different forms before they register and remember it.

LISTEN TO EMPLOYEES

The first and most important step in communicating with employees is to *listen*. No matter how rushed you may be, no matter how bogged down in the everyday details of getting the job done, when an employee has a complaint, concern, or opinion, your priority should be to listen. There are several reasons, all related to improving employee performance, why you should listen to employees:

- *The employee may just be frustrated and in need of an opportunity to vent.* In this case, a few minutes spent listening may get the employee back on track, and his job performance will not suffer. On the other hand, angry or frustrated employees do not work up to par. Therefore, just listening can help maintain and even improve job performance.
- *The employee's input might be evidence of a bigger problem, or the potential for one.* It may be, for example, that a comment by one employee about a working condition is representative of dissatisfaction among a much larger number of employees. In this case, listening might head off employee–management problems.
- *The employee's comments might be a way of pressing a hidden agenda.* A hidden agenda is a secret desire or goal an employee holds but does not admit to. For example, an employee's complaints about a colleague might just be an attempt to cause problems for the colleague. By listening carefully, you can uncover hidden agendas and respond accordingly.
- *The comments might reveal a legitimate need for improvement.* This is particularly true when the comments come from a high-performing employee. When such employees talk about factors that negatively affect their jobs, they present you with an opportunity to make productivity improvements.

CONTINUALLY IMPROVE EMPLOYEE JOB SATISFACTION

Job satisfaction is the foundation upon which productivity growth can be built. Consequently, it is an important element in motivating employees. Supervisors have a key role to play in enhancing the job satisfaction of their direct reports. Factors related to job satisfaction include wages, benefits, working conditions, coworker relationships, the supervisor–employee relationship, potential for advancement, potential for development, new challenges, and competition (see Figure 9.2).

In the area of compensation (wages and benefits) supervisors are typically limited in what they can do other than completing periodic performance appraisals, the results of which can affect an employee's wages. Supervisors can work with employees and higher management to improve working conditions, however.

Supervisors can do a great deal to promote positive relationships among their direct reports. Team-building activities serve this purpose, as does effective conflict management. To promote positive relationships between themselves and their direct reports, supervisors can make a point of being honest, fair, consistent, and regular in their effective communication.

The potential for advancement can also be an effective motivator, particularly when dealing with ambitious employees. However, dangling the "promotion carrot" in front of ambitious employees when there is no hope of a promotion can backfire and kill their motivation. It is better that supervisors be realistic with their employees and, when there is an opportunity, advocate for them when it comes to promotions. Employees who see this happen will be motivated by it.

Development potential is also an effective motivator for ambitious employees. Employees understand that the more job-related skills they develop and the higher the level of those skills, the greater their chances are of moving up. Basing access to cross-training, reimbursement of off-duty education, and other employee development activities on daily job performance can motivate employees to higher levels of performance.

Factors That Promote Job Satisfaction

- Wages
- Benefits
- Working conditions
- Coworker relationships
- Supervisor–employee relationships
- Potential for advancement
- New challenges
- Competition

FIGURE 9.2 Job satisfaction is an important factor in internal customer satisfaction.

New challenges can be used to motivate employees who feel bogged down or bored with their current positions. Sometimes a change of pace is all that is needed to reignite an employee's spark. The opportunity to try something new by taking on a special project, filling in for another team member, or serving on an interesting committee can give employees a new challenge that serves as an effective motivator.

The strategies set forth in this section are aimed at motivating by improving job satisfaction. Increased job satisfaction results in better motivation for some employees. But like all strategies, job satisfaction has its limits. Some employees will be lethargic even when satisfied with their jobs.

EMPOWER EMPLOYEES

Empowerment is one of the most misunderstood elements of the ECS philosophy and one of the most misrepresented by its detractors. The basis for employee empowerment is twofold. First, it increases the likelihood of a good decision, a better plan, or a more effective improvement by bringing more minds to bear on the situation—and not just any minds, but the minds of the people who are closest to the work. Second, it promotes ownership of decisions by involving the people who will have to implement them.

Empowerment means not just involving people, but involving them in ways that give them a real voice. One way this can be done is by structuring work so that employees can make decisions concerning the improvement of work processes within well-specified parameters. Should an employee be allowed to unilaterally drop a vendor if the vendor delivers substandard material? No. The employee should have an avenue for offering her input into the matter, however. Should the employee be allowed to change the way she sets up her machine? Yes, if by doing so she can improve her part of the process without adversely affecting someone else's. Having done so, the next step would be for her to show the innovation to other members of her team so that the improvement she has discovered can be standardized.

Summary

1. Internal customer satisfaction is important because employees tend to treat customers the way they—the employees—are treated by managers and supervisors. In the long run, companies cannot maintain external customer satisfaction unless they also maintain internal customer satisfaction.

2. Leadership is the ability to inspire people to make a total and willing commitment to accomplishing organizational goals. This can be accomplished by: (a) setting a positive example, (b) adopting a participative

leadership style, (c) having a sense of purpose, (d) applying self-discipline, (e) being honest with employees, (f) establishing credibility with employees, (g) showing commitment, (h) being fair and impartial, and (i) taking the blame and sharing the credit.

3. Open and frequent communication with employees is essential to internal customer satisfaction. To promote effective communication, supervisors should remember and apply the following guidelines: (a) do not shoot the messenger; (b) communicate corrective feedback constructively; (c) continually improve communication with employees; (d) select the appropriate communication method; (e) use communication to motivate; and (f) use electronic communication wisely.

4. One of the best ways to promote internal customer satisfaction is to listen to employees' ideas, concerns, opinions, and complaints. The five-step procedure for handling employee input is: listen, investigate, act, report, and follow up.

5. Employee job satisfaction is an important aspect of internal customer satisfaction. Job satisfaction can be continually improved by being attentive to the following factors: wages, benefits, working conditions, coworker relationships, supervisor–employee relationships, potential for advancement, new challenges, and competition.

6. Employee empowerment means not just involving employees, but involving them in ways that give them a real voice. Empowering employees is an effective strategy for promoting internal customer satisfaction.

Key Phrases and Concepts

Appropriate communication method
Be honest with employees
Commitment
Communicate corrective feedback constructively
Continually improve communication
Credibility with employees
Employee job satisfaction
Empower employees
Fair and impartial
Inspiring
Internal customer satisfaction
Leadership
Motivation
Participative leadership style
Positive example
Self-discipline
Sense of purpose
Shoot-the-messenger syndrome
Take the blame, share the credit
Use communication to motivate
Use electronic communication wisely

Review Questions

1. What is internal customer satisfaction, and why is it important?
2. What is meant by the term "leadership"?
3. List and explain several strategies managers and supervisors can use to lead employees.
4. How can managers and supervisors can promote better communication with employees?
5. Why is it so important to listen to the ideas, opinions, concerns, and complaints of employees?
6. Describe the five-step model for handling employee input.
7. List at least five factors that can affect employee job satisfaction.
8. What is meant by the term "employee empowerment"?

ECS APPLIED: DIVERSIFIED TECHNOLOGIES COMPANY MEASURES INTERNAL CUSTOMER SATISFACTION

In the last installment of this case, the three vice presidents of DTC informed the CEO, David Stanley, of various systemic problems that were inhibiting the quality of the company's communication. There were a variety of problems to be dealt with, but of particular concern were the company's automated telephone system and the lack of a uniform message-taking process. One of the solutions put in place was a companywide training program for all employees on the effective use of the telephone.

"The folder in front of you contains the results of our internal customer-satisfaction survey," said David Stanley to open the meeting with his vice presidents. "I asked a human-resources consultant to conduct the survey so that we could obtain results to guide the next step of our ECS implementation." "How did we do, David?" asked Meg Stanfield, vice president for engineering at DTC. "Well, if the survey were a test in college, I'd say we made a B− or a C+. Let's take a look at the results." With this, David Stanley opened his folder, and the three vice presidents followed his lead.

Stanley explained that, as they might have expected, DTC had some strong points and some weak points when it came to internal customer satisfaction. He guided the vice presidents through the survey, explaining the results as he went. The survey asked employees to indicate their attitudes and opinions in several areas, including management/supervisory leadership, communication, willingness of managers/supervisors to listen, job satisfaction, and empowerment.

"The results are mixed, as you can see," said Stanley. "I see a problem," interjected Tim Wang, vice president for manufacturing. "I see several," commented Jake Arthur, vice president for construction. "I don't mean with the company," corrected Wang. "I mean with the survey." "What is the problem, Tim?" asked Stanley. "These are composite results for the company. We need them broken down by division—one set for manufacturing, one for engineering, and one for construction," suggested Wang. "For example, we don't need Meg wasting time on improving job satisfaction if the need is in my area or Jake's."

"Tim makes a good point, David," said Stanfield. "Yes he does," agreed Stanley. "I had the same thought when the consultant showed me the results. Fortunately, getting the results broken down by division won't be a problem. The consultant is preparing division-level reports for each of you now. You'll have them by Friday." Stanley went on to explain that he wanted the vice presidents to use their respective division-level reports to plan for improvements in internal customer satisfaction in their divisions. "Focus on your division's specific weaknesses," said Stanley.

"From the companywide results, it looks like we all have some work to do in the areas of communication, job satisfaction, and empowerment," commented Jake Arthur. "You are right on target, Jake," agreed Stanley. "In fact, if you examine the responses in those categories closely, I think you'll agree that we need to organize some training for our supervisors." "That shouldn't surprise anyone," claimed Meg Stanfield. "Most of our supervisors are technicians who performed well enough in their jobs to be promoted to supervisor. Unfortunately, being a good technician and being a good supervisor is not the same thing."

"I agree," replied Jake Arthur. "It's like taking the best hitter on the baseball team and making him the coach. Hitting requires process skills, but coaching requires people skills. The same is true of being a technician versus being a supervisor." "Good analogy, Jake, and I understand the problem," said David Stanley, holding up his hand to regain the floor. "Now, let me ask each of you a question. You all have engineering degrees, right?" Knowing the answer to his own question, Stanley proceeded without waiting for an answer. "I didn't have even one course in supervision during my undergraduate engineering program," revealed David Stanley. "Did any of you?" Jake Arthur and Tim Wang shook their heads in the negative. Meg Stanfield said, "Not in my undergraduate program, but of course I had several management and supervision courses when I took my Masters Degree in Engineering Management."

Stanley smiled and said, "Meg, you just walked into my trap. I want you to take the lead in putting together a seminar for our supervisors. Make it mandatory. Jake and Tim, I want you to join your learning-deficient CEO and enroll in the seminar along with our supervisory personnel. We can use the training and Meg can use the support. Meg, let's make sure we hit the companywide weaknesses hard. Also, let's offer the seminar in several different formats and at a variety of times to accommodate employee schedules and contract deadlines. Any questions?" Hearing none, David Stanley said, "Let's get to it!"

DISCUSSION CASES

The following cases provide examples of how the various concepts presented in this chapter might play out in actual companies. The cases are provided to prompt discussion, give the reader a feel for the types of problems confronted in the workplace, and reinforce the ECS concept in question.

CASE 9.1 The Phantom Project Manager

Mike Koncar had come to Mar-Tech Corporation with excellent credentials: the right degree, the right experience, and a take-charge attitude. He had looked like just the person to get the XBM-3 project back on track. The previous project manager possessed excellent technical skills, but he was by his own admission not a manager. When it became clear that he could not keep XBM-3 on schedule, he had voluntarily stepped aside as project manager, and Mike Koncar had been brought in.

During the four months Koncar had been on the job, things had gotten a little better. Work on the project was now better organized, and the employees assigned to the project were being used more effectively and more efficiently. There were still problems, however, and XBM-3 was still behind schedule.

"Tom, you've worked on the XBM-3 project from the beginning," said Angela Clark, Mar-Tech's CEO. What's going on with this project? Why can't we get it on track?" Tom Perkins was the most senior employee at Mar-Tech, and Angela Clark often called on him when she needed the wisdom of his years. "Angela, the original problem with XBM-3 was poor organization and inefficient use of resources. You solved that problem by hiring Mike Koncar. He's an excellent organizer and scheduler. "What's the problem then?" asked Clark, with just a touch of exasperation.

"The problem now is poor communication." "What do you mean?" asked Clark. "Well, Koncar is great with the process side of project management, but less so with the people side," answered Perkins. "His only contact with those of us assigned to XBM-3 is by email. It's gotten so bad that our team members refer to Koncar as the phantom project manager. Do you remember that new employee we hired two weeks ago, Angela?" "The one from Georgia Tech?" quizzed Clark. "That's the one. By the way, Angela, as an aside, he's going to be good. But we can talk about that some other time. The point I want to make now is that he has never even met our phantom project manager."

"You've got to be kidding!" blurted out the CEO. "Not at all," answered Perkins with confidence. "I'm not exaggerating, Angela. Mike Koncar sits in his office all day firing off emails, but he is never seen in person." The CEO found it difficult to believe what she was hearing, so Perkins plowed ahead. "Angela, email can be a wonderful tool, but there are some things that are better done in person. I'd be willing to bet that, with just a few team meetings, we could resolve all the problems that are keeping this project behind schedule."

"Thanks, Tom," said Angela Clark. "I'll have a talk with Koncar today—in person. Now, before you go, Tom, let me ask you a question." "Is it going to be the old would-you-be-willing-to-go-back-to-project-management question Angela?" asked Perkins with a knowing smile. "You still won't do it, will you?" remarked Clark with resignation. "Angela, I'm too old and too grouchy to be a project manager again. I don't have the patience anymore, and I don't need the aggravation." "Tom, you are the best project manager Mar-Tech ever had. I know. I worked for you." "Yep, and you are the best engineer I ever worked with, Angela. I'd do just about anything for you. You know that. In fact, I still wish you were my daughter. Do you think I'm too old for adoption proceedings?" Clark smiled at Perkins. The adoption suggestion was a long-running private joke. "But Angela, I really don't want to be a project manager again. I can help you and the company more during my last two years on the job by mentoring our new engineers and by offering advice to my favorite CEO," offered Perkins with a smile. "All right, Mr. Smooth Talker," said Clark with genuine fondness. "Thanks for letting me know what's going on with XBM-3."

Discussion Questions

1. A brief description of Koncar's abilities would read: "Good with processes, bad with people." Have you ever worked with such a person? Explain.
2. In this situation, what would you do to help Mike Koncar improve his people skills?

CASE 9.2 The Supervisor Who Would Not Listen

"How did this happen?" snapped David Atwell, CEO of Atwell Construction Company (ACC). Atwell was speaking to his vice president for operations, Patricia Washington, and he was clearly upset. Before Washington could answer, Atwell slammed his fist on his desk and shouted, "This could bankrupt us! My grandfather started this company. Now I'll be known as the grandson who killed it!" Washington knew better than to interrupt when the CEO was angry. Instead, she sat back and let him vent. Besides, he had good reason to be upset and even afraid, which Washington knew was the real emotion Atwell was feeling.

Two employees were dead—killed in an accident on ACC's biggest jobsite. In addition, the operator of the crane that had overturned and killed them was badly injured. An OSHA inspector was on the way, newspaper reporters were clambering for information, and a television crew had already filmed the accident site and interviewed numerous eyewitnesses. To make matters even worse, the newspapers and television crews had gotten to the shocked families before Atwell could. The wife of the injured crane operator had been taped saying, "Charlie told me something like this was going to happen. He worried about it all the time, but his boss wouldn't listen to him."

When she sensed that Atwell had vented sufficiently, Washington said, "David, this accident is your fault." "What do you mean it's my fault?" snapped the exasperated CEO. "I haven't stepped foot on that jobsite since the groundbreaking. How can this accident be my fault?" "Listen to me, David, because this is important," said Washington in a tone of voice he had never heard her use. "I worked for your grandfather when you were just a child. Then I worked for your father. They were like twins when it came to running this company. Your grandfather and your father listened to what their employees had to say. Both of them knew that the people out there doing the work on our jobsites knew things they didn't know. So they listened, and this company was better because of it. David, you don't listen, and what's worse, you have established a nonlistening culture among our managers and supervisors. They all follow your lead—they don't listen either. That is why this tragedy happened."

When Atwell tried to defend himself, Washington silenced him with a look that could melt ice. "You don't have a clue what happened here, do you David?" Without waiting for a response, Washington pressed her point. "I interviewed our crew members as soon as the ambulance left this morning. I even talked with our injured crane operator while he was lying on the stretcher. He wanted me to know that he had warned his supervisor several times that our crane was not rated for the loads he was being required to lift. His supervisor told him the same thing every time he complained. Do you know what he told him, David?" Atwell mumbled an inaudible response and shook his head. "He told Charlie, our crane operator, to either stop whining and get to work or collect his pay and get lost! As the EMTs carried him away, he kept repeating the same thing: 'Nobody would listen to me, nobody would listen to me.' "

"Then this is the supervisor's fault," offered the CEO, seeing a glimmer of hope. "We can just fire him and claim we have taken responsible corrective action. We can let the supervisor take the fall." "You know something, David? I thought you might say something like that. But you're wrong. This accident is your fault, and I won't stand by and watch you pass the blame off on someone else. I'm going to tell the OSHA inspector that the fault rests with you and the supervisor." Atwell bolted upright in his chair and challenged his vice president. "You wouldn't dare. Not if you value your job!"

"Actually, I do value my job, David. But like your grandfather and your father, I value the lives of our employees even more," replied Washington with finality. "That's why I can't work for you anymore." With that, Patricia Washington handed David Atwell her letter of resignation, ending a brilliant 40-year career with Atwell Construction Company. "By the way, David, I am on my way to your mother's house. She is very upset about the accident and wants to know why it happened and what we are doing to take care of the families of the dead and injured. There might be some significance in the fact that she called me and not you."

Discussion Questions

1. Have you ever worked for a supervisor who would not listen? How did this affect your performance?
2. Analyze this case. How could this tragedy have been prevented?

CASE 9.3 A Tale of Two Engineers

Vicki Martinez and Jan Newton are both project engineers who supervise work teams consisting of 12 people. Martinez and Newton have the same degree from the same university, the same family background, and even the same hobby (tennis). They are the same age. But when it comes to supervising employees, these two engineers, so similar in other ways, are as different as night and day. Newton is brusk and prescriptive in her interaction with employees. She once told Martinez, "I can't believe you waste time asking employees what they think. Employees are paid to work, engineers are paid to think." Martinez, on the other hand, has a well-deserved reputation for not just seeking employee input, but expecting it. She often tells her team members, "Every person on this team has a brain and an opinion. I want you to give me the benefit of your brain and your opinion."

The employees on Newton's team feel repressed and resentful. Those on Martinez's team feel empowered. The difference between the two teams is stark. Martinez's team outperforms Newton's on every evaluative measure the company uses. Newton attributes the difference in performance to employee capabilities—she thinks Martinez has better employees on her team. The facts clearly contradict her contention, however.

Two of the top performers on Martinez's team were once just mediocre performers on Newton's team. These employees shared some interesting insights with the company's human resources director as part of the mandatory evaluation conducted when an employee changes supervisors. The human resources director asked both employees why their performance had improved so drastically after transferring to Martinez's team. This is what they told the director:

1. Whereas Newton held them back and expected them to do only what they were told, Martinez expected them to think and contribute ideas.
2. Whereas Newton's supervision style promoted employee dissention, Martinez's style promoted loyalty. Her team members want to perform well for her.
3. Newton's supervision style made coming to work stressful, but Martinez's style made coming to work exciting, challenging, and fun.

Discussion Questions

1. Who would you rather work for, Martinez or Newton? Why?
2. Should Newton be allowed to supervise employees?

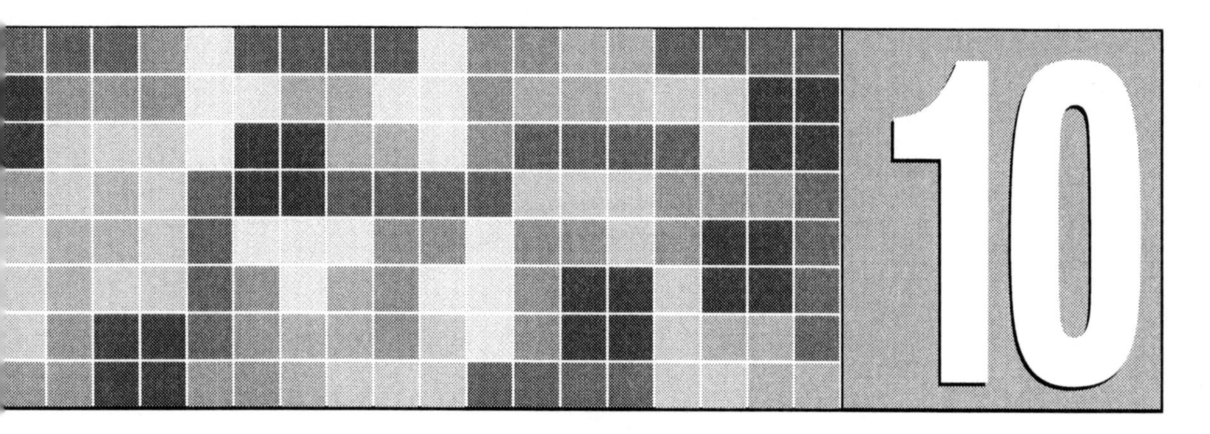

Establish a Customer-Oriented Culture

Special programs and incentives can help promote customer service in the short run, but for the ECS philosophy to take hold for the long run, it must become ingrained in your company's culture. In other words, it must become the normal way of doing business.

GOALS

- Plan for an ECS culture.
- Model ECS behaviors.
- Expect ECS behaviors.
- Monitor and evaluate ECS behaviors.
- Reinforce and reward ECS behaviors.

What does it mean to apply the term "culture" to an engineering, manufacturing, or construction company? The term has been defined in numerous ways as it relates to business, and although most are excellent definitions, we use our own definition.

> *A company's **culture** is the way employees do business when the boss is away. In other words, the culture of a company is the sum of what employees really believe about doing their jobs on a daily basis.*

Another way of viewing a company's culture is that it is the sum of the beliefs that guide behaviors and practices that are reinforced by internal peer pressure. A company's culture determines what is accepted as the normal way of doing business. If employees practice ECS behaviors only when being observed by a supervisor, ECS is not yet a part of the company's culture.

The following strategies can help companies integrate ECS into their corporate culture: (1) plan for an ECS culture (i.e., build ECS into your company's strategic plan); (2) model ECS behaviors; (3) expect ECS behaviors; (4) monitor and evaluate ECS behaviors; and (5) reinforce/reward ECS behaviors.

PLAN FOR AN ECS CULTURE

The ECS approach to doing business is adopted by companies not because it's the nice thing to do, but because it's the right thing to do in terms of good business practices. Stated simply, ECS is good business. Companies that thrive in a competitive marketplace adopt strategies to give themselves a competitive advantage. Smart business leaders do everything they can within a proper ethical framework to ensure successful performance in the marketplace. Establishing an ECS culture is a way to gain a strategic advantage. As such, it should be included in the company's strategic plan. To understand where and how ECS fits into such a plan, one must first understand the various components of a strategic plan, shown in Figure 10.1.

Vision

An organization's guiding force, the dream of what it wants to become, and its reason for being should be apparent in its vision. A vision is like a beacon in the distance toward which the organization is always moving. Everything about the organization—its structure, policies, procedures, and allocation of resources—should support the realization of the vision.

In an organization with a clear vision, it is relatively easy to stay appropriately focused. If a policy does not support the vision, why have it? If a procedure does not support the vision, why adopt it? If an expenditure does not support the vision, why make it? If a position or even a department doesn't support the vision, why keep it? An organization's vision must

Components of a Strategic Plan for Technology Companies

- Vision*
- Mission*
- Guiding principles**
- Broad strategic goals**

*ECS is implied in these components

**ECS is specifically spoken to in these components

FIGURE 10.1 ECS must be built into a company's strategic plan.

be established and articulated by executive management and understood by all employees. The first step in articulating an organizational vision is writing it down.

Writing the Vision Statement

A well-written vision statement, regardless of the type of organization, has the following characteristics:

- It is easily understood by all stakeholders.
- It is briefly stated, yet clear and comprehensive in meaning.
- It is challenging, yet attainable.
- It is lofty, yet tangible.
- It is capable of stirring excitement for all stakeholders.
- It is capable of creating unity of purpose among all stakeholders.
- It is not concerned with numbers.
- It sets the tone for employees.

What follows are examples of corporate vision statements—one for an engineering firm, one for a manufacturing firm, and one for a construction firm. Notice the prominent role of the customer, whether stated or implied, in all three of these vision statements. Also consider the contribution an ECS culture would make to the realization of these visions.

- Altman Engineering Company will be recognized by its customers as the provider of choice in the southeastern United States for mechanical and electrical engineering services.

- Pendleton Manufacturing Company will be recognized by customers as the leading supplier of fireproof storage cabinets in the United States.
- Foster Construction Company will be recognized by customers as the premier road builder in the tri-state region.

Mission

We have just seen that the vision statement describes what an organization would like to be. It's a dream, but it's not "pie in the sky." The vision represents a dream that can come true. The mission takes the next step and describes *who* the organization is, *what* it does, and *where* it is going.

In developing the mission statement for any organization, one should apply the following rules of thumb:

- Describe the who, what, and where of the organization, making sure the "who" component describes the organization and its customers.
- Be brief, but comprehensive. Typically, one paragraph should be sufficient to describe an organization's mission.
- Choose wording that is simple, easy to understand, and descriptive.
- Avoid *how* statements. How the mission will be accomplished is described in the strategies section of the strategic plan.

What follows are corporate mission statements for the three companies whose vision statements were given in the previous section. Notice the importance of the customer, although only implied in this case, in each of these mission statements.

- Altman Engineering Company is a mechanical and electrical engineering firm dedicated to providing research, development, design, planning, and product integration services to an ever-increasing customer base in the southeastern United States.
- Pendleton Manufacturing Company is a hazardous-materials-storage-container producer dedicated to making the work environment of its customers safe and healthy. To this end, PMC produces high-quality, fireproof cabinets for safely storing hazardous materials in an industrial setting.
- Foster Construction Company is a domestic builder dedicated to providing customers with roads and bridges of the highest quality. To this end, FCC provides comprehensive engineering and construction services to customers in the tri-state region.

The missions of these companies are built around serving customers within well-defined markets and geographical regions. Consider the role ECS can play in helping these companies carry out their missions.

Guiding Principles

Guiding principles are written statements that express a company's core beliefs and corporate values. These principles establish the framework within which the company pursues its vision and mission. In a company dedicated to the ECS philosophy, there must be a guiding principle that speaks to customer service or customer satisfaction. What follows are sample guiding principles that might be part of the strategic plan of any engineering, manufacturing, or construction company:

- XYZ Company will uphold the highest ethical standards in all of its operations.
- At XYZ Company, customer satisfaction is the highest priority.
- XYZ Company will make every effort to deliver the highest-quality products and services in the business.
- At XYZ Company, all stakeholders (customers, suppliers, and employees) will be treated as partners.
- At XYZ Company, employee input will be actively sought, carefully considered, and strategically used.
- At XYZ Company, continual improvement of products, processes, and people will be the norm.
- XYZ Company will provide employees with a safe and healthy work environment conducive to consistent peak performance.
- XYZ Company will be a good corporate neighbor in all communities where its facilities are located.
- XYZ Company will take all appropriate steps to protect the environment.

Broad Strategic Goals

Broad strategic goals translate an organization's mission into measurable terms. They represent actual targets towards the achievement of which the organization will expend energy and resources. Broad goals are more specific than the mission, but they are still broad. They still fall into the realm of *what* rather than *how*. Well-written broad organizational goals have the following characteristics:

- They are stated broadly enough that they don't have to be continually rewritten.
- They are stated specifically enough that they are measurable, but not in terms of numbers.
- They are each focused on a single issue or desired outcome.
- They are tied directly to the organization's mission.
- They are in accordance with the organization's guiding principles.
- They show clearly what the organization wants to accomplish.

Broad goals apply to the overall organization, not to individual departments within the organization. In developing its broad goals, an organization should begin with its vision and mission. The broad strategic goals should be written in such a way that their accomplishment will give the organization a sustainable competitive advantage in the marketplace. What follows are examples of broad strategic goals that might be part of the strategic plan of an engineering, manufacturing, or construction company.

- To establish and maintain a world-class workforce at all levels of the organization
- To increase the organization's market share for its existing products/services
- To introduce new products/services to meet emerging needs in the marketplace

With ECS built into the company's strategic plan, it should be apparent to all stakeholders that customer satisfaction is a high priority. We say "should" because, before this can be apparent to stakeholders, they must be shown the strategic plan and be familiarized with its contents. Planning for ECS is a critical first step, but to do any good, the plan must be effectively communicated to all stakeholders.

MODEL ECS BEHAVIORS

In an ECS setting, it is even more important than usual that managers and supervisors set a consistent, positive example of the behaviors they expect of employees. Managers and supervisors must also remember that they have a dual role in modeling the desired ECS behaviors: an external role and an internal role.

If they want employees to listen to customers, managers and supervisors must listen to both internal and external customers. If they want employees to communicate effectively with customers, managers and supervisors must communicate effectively with both internal and external customers. If they want employees to maintain their composure when dealing with difficult customers, managers and supervisors must do so when dealing with both internal and external customers. Detractors will look for any excuse to reject the ECS philosophy (or any other innovation they don't like). Don't let your poor example be their excuse.

EXPECT ECS BEHAVIORS

For ECS to become a cultural imperative in a company, two things must happen. First, all people in positions of authority must expect employees to practice ECS consistently every day. Second, employees must expect each other to practice ECS, so that peer pressure becomes a major enforcer of expectations. There are several ways that managers and supervisors can communicate to employees what they expect. They include the company's strategic plan, employee job descriptions, performance evaluations, the reward/recognition system, and the development and distribution of a customer-commitment statement.

Including customer-service expectations in the company's strategic plan was covered at the beginning of this chapter. Incorporating customer-service expectations in employee job descriptions was covered in earlier chapters. Using performance appraisals and the reward/recognition system to convey and reinforce expectations is covered later in this chapter. The next section deals with developing and distributing the customer-commitment statement.

Customer-Commitment Statement

One of the best ways to demonstrate a commitment is to put it in writing. That is why contracts are always put in writing. When you put your commitment in writing and share it with all stakeholders, you establish both expectations and accountability. It is difficult to give a half-hearted effort to a commitment in this situation. It's like going on a diet and telling everyone you know that you are dieting. Every time you eat, they watch what you eat and how much.

Figure 10.2 is an example of a customer-commitment statement. Several important factors apply to the development, dissemination, and use of such a statement.

- *Developing the commitment statement.* A company's customer-commitment statement should be the product of the input of all stakeholders. This means that the team that develops the statement should include representative managers, supervisors, and employees from all departments in the company as well as customers. Each stakeholder representative should solicit input from his constituency and use it in developing the commitment statement. The company's top executive should approve the final document.

- *Disseminating the commitment statement.* The final, approved commitment statement should be widely disseminated among all stakeholders. We recommend at a minimum the dissemination strategies set forth in Figure 10.3.

- *Using the commitment statement.* A written customer-commitment statement has several potential uses, including the following: (1) it reminds employees of the company's expectations; (2) it helps create peer pressure in support of customer service among employees; (3) it holds the company accountable; (4) it gives customers permission to expect quality customer service; and (5) it serves as an excellent marketing tool.

ABC Company
Customer-Commitment Statement

ABC Company is committed to quality customer service in all of its business transactions. To this end, our interaction with customers will be guided by the following principles:

- Customer satisfaction is our highest priority.
- Customers define the quality of our products and services.
- When customers talk, we listen.
- We actively seek, listen to, and use customer input and feedback.
- Customers are always treated with courtesy and respect.

FIGURE 10.2 Putting your commitment to customer service in writing creates companywide expectations.

Strategies for
Disseminating the Customer-Commitment Statement

- Place framed copies in conspicuous locations throughout company facilities.
- Give a copy to every employee, with a personal letter from the CEO.
- Give a copy to every customer, with a personal letter from the CEO. (Customers should receive a copy as part of their initial contact with the company.)
- Give a copy to potential customers as part of the marketing process.
- Give a copy to suppliers so they understand your company's expectations.

FIGURE 10.3 The customer-commitment statement should be broadly disseminated.

MONITOR AND EVALUATE ECS BEHAVIORS

There is an old management adage that says, "If you want to improve performance, measure it." Measuring performance creates accountability and allows progress to be gauged—like lines on a football field. If a given behavior is expected of employees, the employees should be held accountable for it. This is why companies conduct performance appraisals. Performance appraisals are formal tools that measure employee performance in the areas that are most critical to the company's success. To ensure an effective performance-appraisal process, it is important that companies monitor ECS behaviors daily and evaluate them periodically.

Monitoring ECS Behaviors

Performance appraisals are conducted at regular intervals (every 3, 6, or 12 months, for example), but monitoring should occur on a daily basis. As supervisors interact with their direct reports, they should monitor ECS behaviors in real time. If an employee's behavior falls short, it should be corrected immediately. The customer-commitment statement can be used for pointing out expectations. When supervisors observe a need for training, they should move immediately to arrange that training. Monitoring is an all-day, everyday undertaking.

Evaluating ECS Behaviors

If a company is committed to ECS, there must be at least one customer-service criterion in the company's performance-appraisal instrument. Figure 10.4 contains examples of ECS-related criteria that might be included in performance-appraisal instruments for engineering, manufacturing, and construction companies.

REINFORCE AND REWARD ECS BEHAVIORS

Attempts to institutionalize ECS behaviors will fail unless they include implementation of an appropriate compensation system. Said another way, if you want ECS behaviors to work, make them pay.

The most successful compensation systems combine both individual and team pay. This is important because few employees work exclusively in teams. A typical employee, even in the most team-oriented organization, spends a percentage of her time involved in team participation and a percentage involved in individual activities. Even those who work full-time in teams carry out individual responsibilities on behalf of the team.

Consequently, the most successful compensation systems have the following components: (1) individual compensation, (2) individual incentives,

Examples of ECS Criteria for Performance Appraisal Instruments

- This employee goes the "extra mile" to ensure customer satisfaction.

Always	Usually	Sometimes	Seldom	Never
5	4	3	2	1

- This employee exemplifies the company's customer-service commitment.

Always	Usually	Sometimes	Seldom	Never
5	4	3	2	1

FIGURE 10.4 Performance appraisal instruments should contain at least one criterion relating to customer service.

and (3) team incentives. Under such systems, employees receive their traditional individual base pay. In addition, there are incentives that allow employees to increase their income by surpassing goals set for their individual performance. Finally, the systems include incentives based on team performance. In some cases, the amount of team compensation awarded to individual team members is based on their individual performance within the team; in other words, on the contribution they made to the team's performance.

An example of this approach can be found in the world of professional sports. All baseball players in the National and American Leagues receive a base amount of individual compensation. Most players' contracts also include a number of incentive clauses to promote better individual performance. Team-based incentives are offered if the team wins the league championship or the World Series. When this happens, the players on the team divide the incentive dollars into shares. Every member of the team receives a certain number of shares based on his perceived contribution to the team's success that year.

A four-step model can be used for establishing a compensation system that reinforces both team and individual performance. The first step in this model involves deciding what performance outcomes (for both the individual and the team) will be measured. Step 2 involves determining how the outcomes will be measured. What types of data will tell the story? How can these data be collected? How frequently will the performance measurements be taken? Step 3 involves deciding what types of rewards will be offered—monetary, nonmonetary, or a combination of the two. This is the step in which rewards are organized into levels that correspond to levels of performance so that the reward is in proportion to excellence of performance.

The issue of proportionality is important when designing incentives. If the reward for just barely exceeding a performance goal is the same as the reward for substantially exceeding it, just barely is what the organization will get. If exceeding a goal by 10% results in a 10% bonus, then exceeding it by 20% should result in a 20% bonus, and so on. Proportionality and fairness are characteristics that employees scrutinize with care when examining incentive formulas. Any formula that is seen as unfair or disproportionate will be ineffective.

The final step in the model involves integrating the compensation system with other performance-related processes, such as performance appraisals, promotions, and staffing. If ECS behaviors are important, one or more criteria relating to them should be included in the organization's performance appraisal process.

The employee's ECS criteria ratings in a performance appraisal should be considered when making promotion decisions. A person ineffective with customers should not be promoted in an organization that values ECS. Other employees will perceive this inconsistency, and ECS will be undermined. Finally, during the selection process, applicants should be questioned concerning their views on ECS. It makes no sense for an organization that values ECS to hire new employees who, during their interview, show no interest in or aptitude for ECS.

Nonmonetary Rewards

When organizations develop incentives, they often mistakenly think that employees respond only to dollars in a paycheck. In reality, nonmonetary rewards can be just as effective. Nonmonetary rewards that have proven to be effective include movie or event tickets, gift certificates, time off, free attendance at seminars, getaway weekends for two, airline tickets, and prizes such as electronic or household products.

Different people respond to different incentives. Consequently, it can be difficult to predict what will work best. A good rule of thumb when selecting nonmonetary incentives is: "Don't assume—ask." Employees know what appeals to them. Before investing in nonmonetary incentives, survey your employees. List as many different potential nonmonetary rewards as possible and let employees rate them. In addition, set up the incentive system to include levels of awards to correspond to levels of performance. Then, employees who exceed performance goals by 10 percent, for example, will be allowed to select from among several equally valuable rewards on the "10 percent menu." The better an incentive program is able to respond to individual preferences, the better it will work.

Recognizing and Reinforcing ECS

One of the strongest human motivators is *recognition*. People don't just want to be recognized for their contributions; they *need* to be recognized. The military applies this knowledge very effectively. As we mentioned in Chapter 2, the entire system of military commendations and decorations (medals) is based on the positive human response to recognition. No amount of pay could compel a young soldier to perform the acts of bravery that have been commonplace in the history of the United States military. But the recognition of a grateful nation and grateful comrades-in-arms continues to spur men and women on to incredible acts of valor. There is a lesson here for nonmilitary organizations.

The list of potential methods for recognizing employees goes on *ad infinitum*. Among the many choices, we recommend the following:

- Write a letter to the employee's family, telling about the excellent job the employee is doing.
- Arrange for a senior-level manager to have lunch with the employee.
- Have the CEO of the organization call the employee personally (or stop by in person) to say, "Thanks for a job well done."
- Find out what the employee's hobby is and publicly award him a gift relating to that hobby.
- Designate the best parking space in the lot for the "employee of the month."
- Create a "wall of fame" to honor outstanding performance.

These examples are provided to trigger ideas; they are just a few of the many ways employees can be recognized. Every organization should develop its own locally tailored recognition options. When doing so, the following rules of thumb will be helpful:

- Involve employees in identifying the types of recognition activities to be used. Employees are the best judges of what motivates them.
- Change the list of recognition activities periodically. The same activities used for too long will go stale.
- Have a variety of recognition options for each level of performance. This allows employees to select the type of reward that most appeals to them.

Summary

1. A company's culture is the way employees do business when the boss is away. It is the normal way employees do their jobs and consists of the behaviors that are encouraged by employee peer pressure.

2. Strategies for incorporating ECS behaviors into your company's culture include the following: plan for it, model it, expect it, monitor and evaluate it, and reinforce and reward it.

3. The importance of ECS should be apparent in a company's strategic plan. ECS should appear at least implicitly in the vision and mission statements, and explicitly in the guiding principles and broad strategic goals.

4. Managers and supervisors who expect employees to exhibit desired ECS behaviors must be consistent role models of those behaviors. Few things work more effectively against the cultural inculcation of ECS behaviors than managers and supervisors who tell employees, "Do as I say, not as I do."

5. To incorporate ECS behaviors in an organization, people in authority must expect those behaviors from employees. Employees must also expect them of each other. One of the most effective ways to show all stakeholders what is expected in terms of customer service is to develop and disseminate a customer-commitment statement.

6. Expected ECS behaviors should be monitored daily and formally evaluated periodically. Performance-appraisal instruments should contain at least one ECS-related criterion.

7. Expectations of ECS behaviors should be reinforced frequently using both recognition and rewards. Recognition can range from a pat on the back, to public acknowledgement of a job well done, to some type of nonmonetary perquisite. Rewards range from minor monetary perquisites to cash bonuses and salary increases.

Key Phrases and Concepts

Broad strategic goals
Customer-commitment statement
Evaluating ECS behaviors
Guiding principles
Mission
Monitoring ECS behaviors
Nonmonetary rewards
Vision
Vision statement

Review Questions

1. Define the term "culture" as it applies to engineering, manufacturing, and construction companies.
2. Write an ECS-related guiding principle that could be included in the strategic plan of any engineering, manufacturing, or construction company.
3. Write an ECS-related broad strategic goal that could be included in the strategic plan of any engineering, manufacturing, or construction company.
4. Why is it important for people in positions of authority to be consistent role models of desired ECS behaviors?
5. What are the most important points to remember when developing a customer-commitment statement?
6. To whom should a customer-commitment statement be disseminated?
7. What are the various uses of a customer-commitment statement?
8. Develop an ECS-related criterion that could be included in the performance-appraisal instrument of any engineering, manufacturing, or construction company.
9. How can companies reinforce ECS behaviors?
10. How can companies reward ECS behaviors?

ECS APPLIED: DIVERSIFIED TECHNOLOGIES COMPANY CELEBRATES ECS

In the last installment of this case, DTC's three vice presidents brought the company's CEO up to date on their efforts to identify internal-customer-satisfaction problems. They shared what they were doing to improve internal customer satisfaction in their respective divisions. In this installment, the company's executive management team makes plans to celebrate the progress made to date in the implementation of ECS.

"I want this banquet to be part of our implementation," said David Stanley. "I want it to do several things for us. First, I want it to nail down for all stakeholders that we have implemented ECS, and that it is now the expected way of doing business. Second, I want it to reward each of you and our employees for a job well done. Third, I want it to inform our customers that ECS is supposed to be the norm at DTC, and that I want them to hold us accountable. Finally, I want to use the banquet as a forum to say that, in spite of the excellent results to date, we still have work to do. I want everyone to understand that ECS is a journey, not a destination, and that continual improvement is the goal."

"Good speech, David," said Meg Stanfield, vice president of engineering. "You need to give that same speech at the banquet." "I might do that," responded Stanley, "but first I want to have a couple of our customers speak. We have several customers who have interesting before-and-after stories to tell. I think they will make quite an impression on our employees." "Good idea," agreed Tim Wang, vice president for manufacturing. "Hearing from the customers themselves how we have improved will solidify what we are trying to do better than anything we might say." "Who are the customers you've invited to speak?" asked Jake Arthur, vice president for construction. "They are both customers we had lost but have since won back because of ECS," answered Stanley with a satisfied smile. "But who are they?" prodded Tim Wang. "It's going to be a surprise," said Stanley with a knowing smile. "But believe me, you and everyone else will be impressed with their message."

"What about recognition awards?" asked Meg Stanfield. "Are we going to recognize any of our employees for the parts they played in the ECS implementation or for outstanding customer service?" "I think we should," acknowledged Stanley. "What do my vice presidents think?" "Let's do it," concurred Tim Wang. "We need to say thank you to some of our people." "I agree," offered Stanfield and Arthur in unison. "Why don't you three be the committee to receive nominations and select the recipients?" The three vice presidents nodded their agreement.

"Why don't we ask our customers to nominate employees?" asked Stanfield. "That's an interesting idea," offered Tim Wang. "Yes, and it fits with the ECS philosophy," added Jake Arthur. "Do you mean instead of internal nominations?" asked Stanley. "No," responded Stanfield. "I mean in addition to. In fact, I think we need three categories of awards: the customers' choice award, the company's choice award, and the employees' choice award." "Now that's an excellent idea," interjected Arthur. "The employees' choice award could recognize one of our employees for outstanding *internal* customer service."

"I'm liking all of this better and better," said David Stanley. "What if we give the three awards as Meg explained them, but within each division?" suggested Tim Wang. "I don't want the awards to create hard feelings among the employees in our three divisions." "He's right," offered Jake Arthur. "That means we'll give a total of nine awards, three for each division." "I like that idea too," offered Meg Stanfield. "Me too," said Tim Wang. "That settles it, then," said Stanley. "Why don't the three of you pull together an ad hoc team to plan the banquet and the awards process? I'll take responsibility for finding the money to fund the banquet and whatever you decide to give as awards."

The company's ECS banquet took place two months later and was a resounding success. David Stanley's prediction that everyone would be impressed with the comments of the customers who had speaking roles turned out to be an understatement. Nothing could have had a greater or more positive impact. When the recognition awards were presented, it quickly became clear that the customer-choice awards were an excellent idea. But it was David Stanley who made the greatest impression during the banquet. He had compiled before-ECS and after-ECS earnings figures. Since implementing ECS, DTC had enjoyed a 16.5 percent increase in profits. To make the value of the ECS philosophy personal for every employee, Stanley announced a cash bonus of 2 percent of their base salaries.

DISCUSSION CASES

The following cases provide examples of how the various concepts presented in this chapter might play out in actual companies. The cases are provided to prompt discussion, give the reader a feel for the types of problems confronted in the workplace, and reinforce the ECS concept in question.

CASE 10.1 Good on the Outside, but Bad on the Inside

"What do you mean I need to work on my customer-service skills?" shouted Amy Snow indignantly. "I have better relationships with our customers than anybody in this company!" "You do with some of our customers, Amy," responded Carla Bentley with an apologetic smile. "But not all of them." "I can't believe you are saying this, Carla. You are my best friend, or so I thought." Snow and Bentley are project engineers for the same company. "You should know better than anyone how hard I work on customer service. In fact, I'm the person you and everyone else around here calls when they have a customer who is too hot to handle." "I know, Amy. You are the best we have at handling certain customers," said Bentley. "What do you mean by certain customers?" snapped Amy Snow. "Amy, when it comes to handling customers you are just like that pie I tried to make last week." "Do you mean the one that had a great crust, but was only half-baked on the inside?" teased Amy Snow, happy to be able to turn the tables on her friend. "That's the one," answered Carla Bentley with a sly smile. "It was good on the outside, but bad on the inside."

"You know I hate these little riddles you seem to love so much, Carla, but I'll bite. How am I like that pie?" "It's simple," responded her friend. "You are good, in fact, you are the best at dealing with our external customers. But you are bad when it comes to dealing with our internal customers—our employees. So you see, you are good on the outside, but bad on the inside." "I don't treat employees that badly, do I?" asked Amy Snow. "Actually, it depends on your mood," replied Bentley. "When you're in a good mood, you treat them pretty well—although not as well as you treat external customers." "You mean *real* customers, don't you, Carla?"

"That's the problem, Amy. You don't accept the fact that employees are customers too." "No I don't," snapped Amy Snow decisively. "You don't pay customers, they pay you. So how can an employee be considered a customer?" "Even if you were right, Amy—which you aren't—you are missing the point." "I am, am I? Well then just what is the point, Carla?" "The point is this, Amy," said Carla Bentley with an edge to her voice. "Our employees see you as a role model. So they tend to treat each other and our external customers the way you treat them—the employees. When you snap at employees or brush them off because you're in a hurry, they follow your lead and do the same thing to each other and to our external customers. Believe me Amy, I see this happen all the time." Amy Snow was quiet for a few moments, started to say something, then decided against it.

Carla Bentley sat back quietly and let her friend think. "Has anybody said something to you about this—our boss for instance?" "No, Amy, this is just an observation from a friend. But eventually somebody is going to say something, and I don't want this to hurt your career." Amy Snow was thoughtful and quiet again. When she finally spoke, she said, "I understand, Carla, and I appreciate your concern. I'll work on it." "That's great, Amy!" said Carla Bentley, clearly relieved. Then Amy Snow looked at Carla Bentley with a strange twinkle in her eye and said, "But there is one thing I need to tell you, my friend." Carla Bentley, unsure of what her friend was going to say, whispered to herself, "Uh-oh. Here it comes. I've really offended her." Amy Snow then smiled and said, "The pie analogy was weak, Carla—really lame." "I know," mumbled a relieved Bentley, "but it was the best I could do on short notice."

Discussion Questions

1. Do you agree that employees tend to treat each other the way their supervisors treat them? Explain.
2. How might the way employees and supervisors treat each other affect the way they treat customers?

CASE 10.2 Accountability Is Tough

"I wish we had never put our customer commitment in writing," said an exasperated Nancy Bishop with a sigh. "What's wrong, Nancy?" asked Mark Graison, a colleague of Bishop's at Manufacturing & Engineering Services Company (M & E). "I just met with Rob Carter about the Delta-X job. Rob's our customer contact for Delta-X. You won't believe what he did." "Tell me," said Graison. "He complained that I don't return his calls promptly enough. I told him I've been busy and apologized, but that wasn't good enough for him. He pulled out a copy of our customer-commitment statement and told me I wasn't living up to expectations."

"Ouch," replied Graison. "I believe it," snapped an indignant Bishop. "I felt like rolling that statement up and . . . " "Hold on a minute," interrupted

Graison holding up his hands. "You wouldn't be talking about the customer-commitment statement you helped develop, would you?" Bishop had represented the engineering department on the team that developed M & E's customer-commitment statement. "You know good and well that's the statement I'm talking about. Don't be cute!"

"Nancy, do you remember when I decided to go on a diet several months ago?" asked Graison. "I remember. You told everybody in the company about it." "That's right," acknowledged Graison. "Telling everybody was part of the plan. I did it to establish accountability, and it's working: I've lost 15 pounds." "I know," said Bishop. "You look great." "Thanks," replied Graison with obvious pride. "You helped me." "What did I do?" asked a puzzled Bishop. "Remember that day about a month ago when we met with the CPT team for lunch?" queried Graison. "I remember. I scolded you for ordering dessert. You were so embarrassed you didn't eat it."

"That's right," acknowledged Graison with a knowing smile. "If I hadn't told you and everyone else about my diet, I would have eaten that pie and a lot of other things I don't really need. Accountability is tough Nancy, but without it I never would have lost even the first pound." "All right, all right," said Bishop with resignation. "I get the point." "I'll tell you something, Nancy," remarked Graison. "I've learned that accountability is an unforgiving concept, but it does work." Bishop smiled at her friend and said, "Thanks, Mark. If you'll excuse me, I think I'll go call Rob Carter."

Discussion Questions

1. Do you agree with putting customer commitments in writing? Why?
2. Discuss how the level of accountability increases when commitments are put in writing.

CASE 10.3 You Expect ECS, but You Don't Reinforce It

"The consultant we hired to review our ECS implementation gave us a score of 80 points out of a possible of 100," said Monika Sanders to her executive vice president, Steve Hansen. "Back when we were in college, that would have been a C." Clearly, Sanders was disappointed. Implementing ECS at ATC had been her idea when she took over as CEO of AdvanTech Corporation less than a year ago. The executive management team Sanders inherited when she took over AdvanTech had been skeptical. Hansen had been her only supporter, but Sanders had insisted on the ECS implementation anyway. As Sanders saw it, she had no choice. The owners had fired her predecessor because, under his leadership (or lack of it), AdvanTech had developed a reputation for being antagonistic toward its customers.

As a result, business for AdvanTech had plummeted at a time when it should have increased. Sanders had seen ECS as the best way to change AdvanTech's culture and win back lost customers. And, in fact, ECS was working. Lost customers were coming back—not all, but some of them. The owners were pleased with AdvanTech's performance under Sanders's leadership. Now even the most skeptical of the original detractors had become ECS advocates. All this was why Sanders was so disappointed with a composite score of just 80 points in the consultant's final report. She had been sure the final score would be at least 95 and hoped it would be even higher.

"Monika, you are taking this score a little too hard," said Hansen as he flipped through the final report. "Did you read the consultant's executive summary? It's actually very positive." "An 80 out of 100 is not positive, Steve. I had hoped to use this report to show off a little to the owners and to give a pat on the back to employees." "Maybe we can still do that," offered Hansen. "Monika, look at how the consultant organized his report. It has five parts: planning, role modeling, expectations, monitoring and evaluation, and reinforcing and rewarding. Each part has a possible maximum of 20 points. Look at the points for each part." The consultant's summary contained a chart with the following scores:

Planning	20 points
Modeling	15 points
Expectations	15 points
Monitoring and evaluating	15 points
Reinforcing and rewarding	15 points

"Monika, what this report really says is that we have done an excellent job so far in our implementation of ECS, but that we still have some work to do. I suppose there is a sense in which an 80 at this point is actually better than a 100," said Hansen. Sanders simply raised an eyebrow and made a gesture that said, "Tell me more." "If we had gotten a perfect score, our people might have been tempted to relax and think there was nothing more to do." "I've worried about that a little. I don't want to lose momentum," suggested Sanders.

"Good point," said Hansen. "I like the idea of telling employees that we've bought the house, so to speak; now we need to paint it, put in new carpet, and plant some grass." "That's not a bad analogy, Steve. Why don't we get our managers and supervisors together, review the consultant's report, and develop an action plan for improving the 15s to 20s. Then, when we share the report with the owners and employees, we can also give them the plan." "Now you're talking, Monika," said an enthusiastic Hansen. "I like that approach."

"All right, Steve, one more thing—let's talk about perceptions. I don't want this consultant's report to be perceived as a grade of C. Do you have any ideas?" "Well, there is a fairly simple solution to the perception problem," offered Hansen. "We can use a baseball analogy." "Baseball?" asked Sanders. "Yes. In baseball, a batting average of 80 out of 100 is an .800 average. You can make the all-star team by batting just .350. The best hitter who ever played the game didn't even come close to batting .800."

"I like the analogy," smiled Sanders. "Steve, you're a genius. We'll use the baseball analogy every time we talk about the report. In fact, let's prepare a PowerPoint presentation for our employees that begins with an animated baseball player hitting a home run. Under the baseball player, we will have a caption that says we are batting .800." "Good idea, Monika. I like the subliminal message that says we have hit a home run."

As the meeting broke up, Monika Sanders said, "Thanks for listening, Steve—and thanks again for the baseball idea." "You're welcome, Monika, and thank you. We never would have turned this company around without your commitment to ECS." "Thanks, Steve. By the way, didn't you play baseball in college?" Hansen just nodded. "What was your batting average?" asked Sanders. "I don't want to talk about it," shrugged Hansen, obviously embarrassed. "Come on, Steve; you can tell me. What was your batting average?" "Let me put it this way, Monika. First, it was not anywhere near .800, and second, it's why I'm an engineer instead of a New York Yankee."

Discussion Questions

1. Analyze the consultant's report in this case. If you were CEO of ATC, what would you do first?
2. Do you agree with Sanders's decision to use the batting-average analogy to soften the blow of the consultant's score? Why?

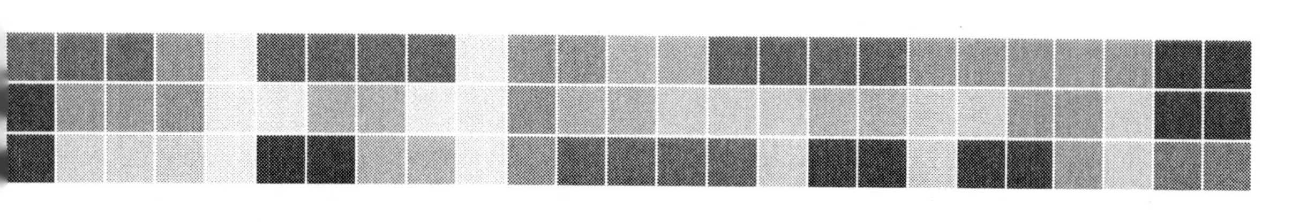

Index